Past Masters

General Editor Kei

# Frege

**Joan Weiner** is Professor of Philosophy at the University of Wisconsin-Milwaukee, where she teaches philosophy of language, logic, and the history of early analytic philosophy. She is the author of *Frege in Perspective* (1990).

# Past Masters

AQUINAS Anthony Kenny
ARISTOTLE Jonathan Barnes
AUGUSTINE Henry Chadwick
THE BUDDHA Michael Carrithers
CLAUSEWITZ Michael Howard
DARWIN Jonathan Howard
DESCARTES Tom Sorell
FREGE Joan Weiner
FREUD Anthony Storr
GALILEO Stillman Drake
GANDHI Bhikhu Parekh
GOETHE T. J. Reed
HEGEL Peter Singer
HEIDEGGER Michael Inwood
HOBBES Richard Tuck
HUME A. J. Ayer
JESUS Humphrey Carpenter
JUNG Anthony Stevens
KANT Roger Scruton
KIERKEGAARD Patrick Gardiner
LOCKE John Dunn
MACHIAVELLI Quentin Skinner
MARX Peter Singer
MUHAMMAD Michael Cook
NIETZSCHE Michael Tanner
PAUL E. P. Sanders
PLATO R. M. Hare
ROUSSEAU Robert Wokler
RUSSELL A. C. Grayling
SCHOPENHAUER Christopher Janaway
SHAKESPEARE Germaine Greer
SOCRATES C. C. W. Taylor
SPINOZA Roger Scruton
WITTGENSTEIN A. C. Grayling

# Frege

## Joan Weiner

OXFORD
UNIVERSITY PRESS

Great Clarendon Street, Oxford OX2 6DP

Oxford University Press is a department of the University of Oxford.
It furthers the University's objective of excellence in research, scholarship,
and education by publishing worldwide in

Oxford New York

Athens Auckland Bangkok Bogotá Buenos Aires Calcutta
Cape Town Chennai Dar es Salaam Delhi Florence Hong Kong Istanbul
Karachi Kuala Lumpur Madrid Melbourne Mexico City Mumbai
Nairobi Paris São Paulo Singapore Taipei Tokyo Toronto Warsaw

with associated companies in Berlin Ibadan

Oxford is a registered trade mark of Oxford University Press
in the UK and in certain other countries



First published as an Oxford University Press paperback 1999



British Library Cataloguing in Publication Data

Library of Congress Cataloging-in-Publication Data
Data available

Weiner, Joan.
Frege / Joan Weiner.
Includes bibliographical references and index.
1. Frege, Gottlob, 1848–1925. I. Title.
B3245.F24 W44 1999 193–dc21 99–32366

ISBN 0–19–287695–3 (pbk.)

10 9 8 7 6 5 4 3 2 1

Typeset by RefineCatch Limited, Bungay, Suffolk
Printed in Great Britain by
Cox & Wyman Ltd.,
Reading, Berkshire

# Preface

Gottlob Frege's writings have had a profound influence on contemporary thought. His revolutionary new logic was the origin of modern mathematical logic—a field of import not only to abstract mathematics, but also to computer science and philosophy. Frege's new logic, however, is only one important part of his contribution to contemporary philosophical thought. Nearly his entire career was devoted to an attempt to determine the nature of our knowledge of the truths of arithmetic. And, although many of the ideas and insights developed in the pursuit of this project remain influential today, perhaps most influential of all was Frege's conception of the project itself. Frege's writings are widely regarded today as the origin of analytic philosophy, the approach that is dominant among Anglo-American philosophers today.

This short introduction to Frege's thought is designed to be accessible to those without background in philosophy, logic, or mathematics. In order to achieve this end, I have had to make some compromises. Due to limitations of space, some very interesting and important writings (including *Logical Investigations*, Frege's last series of papers) have received only cursory discussion. In order to avoid very technical discussions, I have included very little discussion of some of the more sophisticated technical devices that are presented in *Basic Laws*. Finally, it is worth noting that there is a great deal of controversy about how to read Frege's writings. I have tried to indicate areas of controversy in the text. However, there is no way, in a short introduction of this sort, either to introduce the reader to the many different views about how his work should be read or to offer any satisfying defence of the particular reading that is offered. The suggestions for further reading that are included later in this volume should help the reader fill in some of these gaps.

As the above comments suggest, my decision was not to attempt to provide a comprehensive account of Frege's ideas and insights. Instead, I chose to tell the story of the progress of Frege's project. This story begins with the motivation and conception of the project; continues with the modifications of the logical system and underlying philosophical views that were necessary to accommodate the demands of the project. It culminates with Russell's discovery of a fundamental contradiction in Frege's system and with Frege's post-contradiction reflections. This is, in my view, one of the most compelling and exciting stories in the history of philosophy. It is my hope that I have managed to convey some of this to the reader.

I would like to take this opportunity to thank Rebecca Hunt who first contacted me and asked me to consider writing a Frege volume for the Past Masters series. Shelley Cox, who took over when Rebecca Hunt left the Press, has been helpful at every stage and provided me with particularly useful comments on the first draft of the manuscript. I would also like to thank an anonymous reader for the Press for helpful criticisms and suggestions. Some of the details of Frege's biography come from Werner Stelzner's *Gottlob Frege: Jena und die Geburt der modernen Logik* (Verein zur Regionalförderung von Forschung, Innovation und Technologie für die Strukturentwicklung e.V.,1996), and Lothar Kreisel's 'Alfred' in *Frege in Jena* (Königshausen & Neumann, 1997). I am also indebted to Gottfried Gabriel and Uwe Dathe for discussions and tours of the Frege sights in Jena. Most of all, I thank Mark Kaplan, for his detailed and invaluable comments and criticisms; for endless discussion; and for his unfailing faith in the project. This book is dedicated to him with love, gratitude, and an appreciation of my extraordinary luck.

# Contents

# Note on references

The following is a list of abbreviations, along with the titles in published English translations, that I have used in the citations of Frege's writings. Frege's writings are now available in a variety of published volumes in English and in German. Nearly all these volumes include either section numbers or the page numbers of the original publications in the margins, for writings that were published in Frege's time. Thus my citations of these works are to section numbers or the page numbers on which the passage appeared in the original publication. Page numbers that appear in citations of writings that were not published in Frege's time are to two volumes, the volume in which these writings are published in German (*Nachgelassene Schriften*) and the volume in which they are translated into English (*Posthumous Writings*).

| | |
|---|---|
| *B* | *Begriffsschrift* |
| *BLA* | *Basic Laws of Arithmetic* |
| CN | 'On the Scientific Justification of a Conceptual Notation' |
| CO | 'On Concept and Object' |
| *FA* | *Foundations of Arithmetic* |
| FC | 'Function and Concept' |
| MC | 'Methods of Calculation based on an Extension of the Concept of Quantity' |
| *NS* | *Nachgelassene Schriften* |
| *FG* | *On the Foundations of Geometry* |
| *PW* | *Posthumous Writings* |
| SM | 'On Sense and Meaning' |
| T | 'The Thought' |

'*Begriffsschrift*' is both the title of a monograph and Frege's name for his logical notation. In the text that follows, this expression is used without italicization to refer to logical notation and in italics to refer to the monograph.

# 1 *Life and character*

Gottlob Frege (1848–1925) was born in Wismar, a German port town on the Baltic coast. In 1866, after the death of his father, who was the owner and headmaster of a private school for girls, Frege's mother, who had been a teacher at the school, took over the running of the school. Frege began his university studies at Jena, where he took courses in mathematics, physics, chemistry, and philosophy. After studying for two years at Göttingen, where he received his doctoral degree, Frege returned to Jena and wrote his Habilitationsschrift, a post-doctoral thesis required for university teaching. Immediately after finishing his Habilitationsschrift, he was appointed Privatdozent, an unpaid teaching position. His mother sold the school in Wismar and moved to Jena to support and live with her son.

Five years later, in 1879, Frege published the first contribution to the project that was to occupy most of his career: his attempt to show that the truths of arithmetic could be derived from logic alone. In this monograph, *Begriffsschrift,* Frege introduced a revolutionary new logic. As a result of the publication of this work, Frege was promoted to the position of Ausserordentlich Professor, a position that carried with it an increase in prestige and a modest stipend. During this period in his life Frege took frequent hiking vacations in the area in which he had grown up. On one of these trips, he met Margarete Lieseberg. They were married in 1887.

Although *Begriffsschrift* is recognized today as the origin of mathematical logic, its significance was not immediately apparent to Frege's contemporaries. His next great work, *Foundations of Arithmetic,* which was published in 1884, also attracted very little interest. Further progress on the project was delayed for two reasons. One of these was, Frege wrote,

[T]he discouragement that overcame me at times because of the cool reception—or more accurately, the lack of reception—accorded by mathematicians to the writings of mine that I have mentioned. (*BLA*, p. xi)

Another reason for the delay was that Frege discovered difficulties with his original conception of the logic—difficulties that required him to make basic alterations in the mechanics of the logic. As a result, he found that he needed to discard a nearly complete manuscript and begin again. The first volume of *Basic Laws of Arithmetic*, the work that was to have completed Frege's project, was published in 1893.

In 1896 Frege was promoted to honorary full professor. The promotion was a direct consequence of the recognition that Frege's work had finally begun to receive. As a result of favourable mentions by such eminent mathematicians as Giuseppe Peano (1858–1932) and Richard Dedekind (1831–1916), his work had attracted a number of readers, among whom was the English philosopher, Bertrand Russell (1872–1970). In 1902, when the second volume of *Basic Laws* was in press, Russell sent Frege a now-famous letter showing that the logic of *Basic Laws* was inconsistent. Frege went ahead with the printing of the second volume, adding an appendix in which he discussed the contradiction and strategies for avoiding it.

Frege's wife died one year later, after a long illness. Although Frege and his wife had had no children, after her death Frege took responsibility for bringing up a child. In 1908, 5-year-old Alfred Fuchs's mother was seriously ill and his father had been committed to an asylum. No suitable guardian could be found among the relatives of his parents, and the people who knew Alfred in Gniebsdorf regarded him as incorrigible. At the suggestion of Frege's nephew, who was a pastor in Gniebsdorf, Frege became Alfred's guardian. Later, when Alfred came of age, Frege adopted him. Frege was, by all accounts, a kind and loving father. Alfred's school records indicate that he was well behaved and diligent. Alfred ultimately became a mechanical engineer.

It is difficult to fail to be moved by the generosity of spirit suggested by this information about Frege's later life. But it is also difficult to fail to be moved by other features of Frege's character that are not admirable at all. Among his later writings is a diary, mostly about political topics, written in the year before his death. Frege had regarded himself a liberal earlier, but his views changed as a result of the consequences of Germany's loss of the First World War and, in particular, the harsh terms imposed by the Treaty of Versailles. The diary entries of 1924 reveal Frege to have held extreme anti-democratic views and, although the diary contains only a few brief remarks about Jews, these remarks reveal a notable anti-Semitism. Frege deplored the influence of Jews in the National Liberal Party and the influence of Jewish business practices. The longest of these comments is the most chilling. Frege wrote,

> One can acknowledge that there are Jews of the highest respectability, and yet regard it as a misfortune that there are so many Jews in Germany, and that they have complete equality of political rights with citizens of Aryan descent; but how little is achieved by the wish that the Jews in Germany should lose their political rights or better yet vanish from Germany. If one wanted laws passed to remedy these evils, the first question to be answered would be: how can one distinguish Jews from non-Jews for certain? That may have been relatively easy 60 years ago. Now, it appears to me to be quite difficult. Perhaps one must be satisfied with fighting the ways of thinking which show up in the activities of the Jews and are so harmful, and to punish exactly these activities with the loss of civil rights and to make the achievement of civil rights more difficult. (30 Apr. 1924; trans. Richard L. Mendelsohn, ed. with commentary by Gottfried Gabriel and Wolfgang Kienzler, in *Inquiry*, 39 (1996))

As this brief discussion indicates, the available evidence leaves us with a complex picture of Frege's character—a picture that combines admirable and abhorrent features.

The picture of Frege's work on his central project, however, is very different. Frege devoted nearly his entire career to a grand

and beautiful project that combined philosophical and mathematical argument. He continued to push forward in spite of years of discouraging responses to his work. And when, after many years of work, Frege finally produced the work in which he believed he had brought his project to fruition, he was confronted with definitive failure. Some years after his discovery of the contradiction, Bertrand Russell wrote,

As I think about acts of integrity and grace, I realize that there is nothing in my knowledge to compare with Frege's dedication to truth. His entire life's work was on the verge of completion, much of his work had been ignored to the benefit of men infinitely less capable, his second volume was about to be published, and upon finding that his fundamental assumption was in error, he responded with intellectual pleasure clearly submerging any feelings of disappointment. It was almost superhuman and a telling indication of that of which men are capable if their dedication is to creative work and knowledge instead of cruder efforts to dominate and be known. (Jean van Heijenoort (ed.), *From Frege to Gödel: A Source Book in Mathematical Logic 1879–1931* (Harvard University Press, 1967), 127)

Frege published little more in the remaining twenty-two years of his life. Many have assumed that he succumbed to the discouragement that haunted him earlier in his career. The evidence, however, suggests otherwise. Frege worked for some time on finding a solution to Russell's paradox, but ultimately concluded that it could not be solved. His meagre publication record is partly explicable by his having spent most of his efforts on the failed attempt to find a solution. Moreover, he did not abandon his intellectual work after concluding that his original project could not be carried out. For he had already come to this conclusion by 1918, a year in which he wrote, 'In these difficult times I seek consolation in scientific work. I am trying to bring in the harvest of my life so it will not be lost.' (Frege to Hugo Dingler 17 Nov. 1918). The work to which he refers was begun in a series of papers entitled *Logical Investigations*. His aim was to provide a new, informal introduction to his conception of logic. Nor did he give up on his interest in the foundations of arithmetic. In another letter written after 1918, Frege wrote,

> As you probably know, I have made many efforts to get clear about what we mean by the word 'number'. Perhaps you also know that these efforts seem to have been a complete failure. This has acted as a constant stimulus which would not let the question rest inside me. It continued to operate in me even though I had officially given up my efforts in the matter. And to my own surprise, this work, which went on in me independently of my will, suddenly cast a full light over the question. (Frege to Zsigmondy; undated, but after 1918)

In 1925, only three months before his death, Frege was corresponding with the editor of a monograph series about publishing a new account of the sources of our knowledge of arithmetic.

Of course, the story of Frege's dogged determination in the face of failure is not the whole story. If it were, there would be no interest in a book on his work. The most moving and fascinating story to be told about Frege is not a story about a man at all, but a story about a philosophical project. For, while Frege was unable to produce a solution to the problem he set himself, the work he did in the service of this project has left us with a vast and important intellectual legacy. Frege's logic is an important part of this legacy. The consistent part of it formed the basis of modern logic as we know it today—an advance that has been important, not only to philosophers, but also to mathematicians and computer scientists. Equally important, to philosophers, is Frege's conception of his project and the insights that appear in his contributions to this project. These insights have had a profound and lasting impact on contemporary philosophical thought about logic, mathematics, and language.

# 2 *The project*

Frege begins *Foundations of Arithmetic* with a discussion of the question 'What is the number one?' As he acknowledges, most people will feel that this question has already been adequately answered in elementary textbooks. Yet he claims not only that the apparent answers in elementary textbooks are inadequate, but that even mathematicians have no satisfactory answer to offer. Moreover, he continues, if we cannot say what the number one is, there is small hope that we will be able to say what number is. He writes,

> If a concept fundamental to a mighty science gives rise to difficulties, then it is surely an imperative task to investigate it more closely until those difficulties are overcome. (*FA*, p. ii)

But what are the difficulties to which the concept of number gives rise? Frege does not think that difficulties arise for most of us in our everyday use of arithmetic, nor does he think that difficulties impede the work of most mathematicians. The difficulty, on Frege's view, is that even the greatest of mathematicians lack insight into the foundation of the whole structure of arithmetic. If we had such insight we would be able to explain, among other things, the special status of our knowledge of the truths of arithmetic.

Why think that our knowledge of arithmetic has a special status? One reason is that there seems to be a difference between the sort of evidence required to establish the truths of arithmetic and the sort of evidence required to establish most other truths. Everyday knowledge is established by observation; by using evidence of the senses. In order to determine whether there is milk in the refrigerator, I will look in the refrigerator. In order to determine whether the milk in the refrigerator has spoiled, I will smell it. In this respect, everyday knowledge and our knowledge of truths of the physical sciences seem similar. Although it is more difficult to establish most scientific truths than to determine whether there is milk in the refrigerator—we cannot simply look and see, for example, that a particular virus causes a newly recognized disease—nonetheless, evidence of the senses is required. The very recognition of the new disease will be based on evidence of the senses. For example, AIDS was recognized as a new disease because in 1981 people began to show up in hospitals with unusual (and observable) symptoms. The subsequent work that led to the conclusion that a particular virus, HIV, causes AIDS required further observations. In contrast, no evidence of the senses seems to be required to establish that, for example, there is no greatest prime number. It suffices to offer a proof that there is a method that, given any prime number, allows us to show that a larger prime number exists. These sorts of considerations convinced Frege that the source of our knowledge of the truths of arithmetic is different from the source of our knowledge of everyday truths and truths of the physical sciences. His project was to identify the source of our knowledge of arithmetic.

This project requires not simply an appreciation of mathematics and how its truths are established, but also a general view about knowledge and its sources. When Frege began his work he was aware of two available accounts of the sources of knowledge, both of which he found unsatisfactory. The simplest account, and the one for which he had least sympathy, was the empiricist account. On the empiricist account, sense experience is the source of all our knowledge, including our knowledge of the truths of arithmetic. In one respect this account seems correct. Were we to investigate the processes by which we come to believe truths, it is likely that we would find that evidence of the senses is always involved at some point. After all, even a sophisticated proof of number theory will include appeals to elementary truths of arithmetic—truths that we originally learned as small children. And these are typically learned by using evidence of the senses. A small child may learn, for example, that 2+2=3+1 by arranging and rearranging four small objects or she may be taught to memorize it in a classroom. In either case use of the senses is involved.

But an investigation of the sources of knowledge, as Frege understands it, has very little to do with how we actually come to believe truths. For our actual reasons for coming to believe the truths of mathematics might be bad reasons. They might be reasons that do not provide adequate justification of the truths; reasons that do not establish these truths. Many beliefs, even some true beliefs, are based on superstition, but superstition is not a source of knowledge. The source of our knowledge of a truth is determined not by how we came to believe it, but rather by what it is that we actually use to establish or justify it. Frege agreed with the empiricist assessment of our knowledge of truths of the physical sciences. He believed that this knowledge is a posteriori; that is, that it is established by appeals to evidence of the senses. But he disagreed with the empiricist assessment of our knowledge of truths of mathematics. This knowledge, Frege believed, can be established without appeals to evidence of the senses; he believed, that is, that this knowledge is a priori.

Frege's conviction that our knowledge of mathematics is a priori appears to be supported by the difference just sketched between what is required to establish truths of the physical sciences and what is required to establish truths of mathematics. Let us look more closely at this difference. Consider, first, the sort of evidence required by the researcher who is attempting to determine whether a particular virus causes some disease. She cannot come to her conclusions simply by engaging in abstract thought. She will need to carry out tests. A virus that is the cause of a disease must be present in the people suffering from the disease. Thus one part of the task is to develop a means for testing for the presence of the virus—a means for finding visible evidence of the presence of the virus. Another part of the task is to carry out this test on a number of sufferers of the disease. Supposing our researcher comes to the conclusion that this virus does indeed cause the disease, this conclusion will go beyond what she directly observes. Nonetheless, it is her observations that provide the justification for her conclusion. She will need to provide an argument based on her test results and to appeal, in her argument, to her observations; to evidence of the senses. It is for this reason that, on Frege's view, such knowledge has sense experience as its source.

In contrast, arguments used to establish general truths about numbers appear to require no appeals to evidence of the senses. To see this, let us consider an example: an argument that, for any two whole numbers $x$ and $y$, if both are divisible by 5, then their sum is divisible by 5. We begin by exploiting the definition of 'is divisible by'. To say that $x$ is divisible by 5 is (by definition) to say that there is some whole number, say $a$, such that $x=5a$. Similarly, to say that $y$ is divisible by 5, is to say that there is some whole number, say $b$, such that $y=5b$. Hence $x+y=5a+5b$. Since, by the distributive law, $5a+5b=5(a+b)$, it follows that $x+y=5(a+b)$. $(a+\mathrm{b})$ is the sum of two whole numbers. Since the sum of two whole numbers is a whole number, $x+y$ is divisible by 5. This argument contains explicit appeals to the definition of 'is divisible by', to a law of arithmetic (the distributive law), and to a general claim about addition (that the sum of two whole numbers is a whole number). There are also some implicit appeals to laws of identity. Since there are no obvious appeals to evidence of the senses, the argument seems to show that the truth can be known a priori.

One might be tempted to think, however, that there really is an appeal to evidence of the senses. The argument is written out and the reader of the argument needs to look at it (to make use of visual evidence) to read it. In this way, one might suppose, even a mathematical argument relies on evidence from sense experience. But this is misleading. For the visual evidence (the observation of a sentence in a written proof) is not evidence for the conclusion. It is the truth of the sentence that provides evidence for the conclusion. And the appearance of the sentence does not provide evidence of its truth. Indeed the justification of the statement about numbers does not require any observation of written sentences, for the proof need not be written to be convincing. After all, we can figure out a proof without writing it down. There is no obvious need for an appeal to evidence of the senses.

The contrast between the justification of general statements about the physical world and that of general statements about numbers may, nonetheless, be more apparent than real. After all, the above argument for a general claim about numbers includes several appeals that have not yet been examined. Among these is an appeal to the distributive law of arithmetic. If the distributive law can only be established by some sort of justification that exploits the evidence of the senses, then the original truth cannot really be known a priori. In that case, the full (or ultimate) justification of the original truth requires, at some point, evidence of the senses. Thus, in order to determine whether the truths of arithmetic can be known a priori, it is important to look for their full justification, to derive them from the primitive, unprovable truths on which they depend. This is Frege's project.

He was convinced, before beginning the project, that it would result in a demonstration that truths of mathematics really could be known a priori; that no examination of the full justification of these truths would, at any point, reveal a necessary appeal to sense experience. In this, he was not alone. For the empiricist account of knowledge was not universally accepted in Frege's time. The other account that Frege considered has its roots in the work of Immanuel Kant (1724–1804). The Kantian account requires the recognition of another source of knowledge, pure intuition.

On Frege's view this account constituted a substantive advance—both for our understanding of the sources of knowledge in general and for our understanding of the sources of some mathematical knowledge. An important part of this advance was Kant's formulation of the analytic/synthetic distinction. Kant characterized analytic truths as those in which the predicate concept is contained in the subject concept. Analytic truth is what we get through analysis of concepts. For example, to be a grandmother is to be a female grandparent. Assuming that this simply describes the content of the concept *grandmother*, it is an analytic truth that all grandmothers are grandparents. Synthetic truths are those that we cannot get through analysis of concepts. For example, it may be true that all grandmothers are more than 30 years old. But, since being over 30 is no part of the content of the concept *grandmother*, this truth, if it is one, is synthetic. It can only be established by appeal to facts (perhaps facts about human reproduction or perhaps simply the results of an exhaustive survey of all grandmothers). In order to establish a synthetic truth we need to appeal to something beyond the content of the concepts involved. This formulation of the analytic/synthetic distinction enabled Kant, Frege claims, to reveal the true nature of the truths of Euclidean geometry—that these truths are both synthetic and a priori.

What, on this view, is the source of our knowledge of Euclidean geometry? A synthetic truth cannot be established without appealing to something other than the content of the concepts involved. If the truths of geometry are not only synthetic but a priori then, whatever appeals are necessary, they cannot be appeals to evidence of the senses. But what other evidence is available? In order to answer this question, we will need to consider how truths of Euclidean geometry are established.

As with the proof we considered earlier—the proof that the sum of two numbers divisible by 5 is, itself, divisible by 5—the justification of a truth of Euclidean geometry seems to require no appeals to particular observations. What is required to justify a truth of Euclidean geometry is a formal proof from axioms. However, as we have already seen, the fact that a truth of Euclidean geometry is established by a proof from axioms does not, on its own, suffice to show that it is an a priori truth. For the justification of the axioms on which the proof depends is part of its ultimate justification. If the justification of the axioms requires an appeal to particular observations, then the ultimate justification of truths proved from the axioms require such appeals. How, then, are the axioms of Euclidean geometry justified?

Frege claims that anyone who understands the words 'point', 'line', etc. as he does cannot but accept these axioms. One might suppose, then, that the axioms will be derivable from definitions of the terms. But, since the terms in question are primitive, no definitions are available. Frege suggests that the meanings of such primitive terms can be communicated only by hints and figurative modes of expression (*FG* 301). Some examples of hints that are used to communicate the meanings of primitive terms of geometry by Euclid are: a point is that which has no part; a line is breadthless length. Although these hints provide descriptions of points and lines, they are not definitions. Nor can they be used to prove the axioms. The axioms, Frege says, are self-evident. His suggestion is that, having been exposed to such descriptions, we will simply see that the axioms are true.

This account of the justification of the axioms of geometry, however, may seem to contradict the experience most of us have had when we were taught Euclidean geometry. Consider, for example, a particular axiom: Any two points determine a unique straight line. A teacher is not likely to trust the sort of descriptions of points and lines mentioned above to do the work. More likely, the teacher will try to convince a student by drawing a pair of dots and showing how to draw a line between them using a straightedge. Thus it may seem that the axiom really is justified by appeals to particular observations, in just the way that general truths about the physical world are justified by appeals to particular observations.

But, in fact, the role played by particular observations in geometry is very different from the role played by particular observations in establishing general truths about the physical world. Consider the role played by particular observations in the attempt, for example, to show that all people suffering from AIDS have been infected by HIV. The evidence for this hypothesis consists of particular observations of individual positive test results. But the evidence provided by one positive test result (or two, or three) is unlikely to convince anyone of the truth of the hypothesis about the virus. It is important to perform a large number of tests. The researcher who wants to support this hypothesis must include, in her argument, reports of the numbers of test results. More tests constitute better support. In contrast, the geometry teacher does not report on the number of times that the 'experiment' with the straightedge has been carried out. And the geometry teacher is likely to be impatient with a student who would like to see this experiment repeated several hundred times. What explains this difference?

On Frege's view, the answer is clear: these drawings do not constitute evidence for the truth of the axioms. The source of the justification of the axioms of geometry is not sense experience but, rather, pure intuition—a faculty underlying our perception of objects in space. We do not learn about spatial relations by observing physical objects; rather, we cannot observe physical objects *except* as objects in space. It is the spatial structure of perception, not characteristics of particular objects that make it impossible, for example, to perceive three objects, each of which is between the other two. The truths of geometry are general truths that apply to all spatial objects. The drawings are useful for the study of geometry because they are partly idealized illustrations of spatial relations. They get us to recognize what we already understand—certain unalterable features of spatial relations. Rather than constituting evidence for an axiom, a drawing is of psychological help in getting us to see that the axiom is true. Frege remarks,

> I do not mean in the least to deny that without sense impressions we should be as stupid as stones, and should know nothing either of numbers or of anything else; but this psychological proposition is not of the slightest concern to us here. (*FA* §105)

The axioms, once we understand them, are self-evident because an understanding of them is, in some sense, already written into our ability to perceive anything at all. Thus, any arrangements or configurations of physical objects that we can picture, including, Frege says,

> [t]he wildest visions of delirium, the boldest inventions of legend and poetry, where animals speak and stars stand still, where men are turned to stone and trees turn into men, where the drowning haul themselves up out of swamps by their own topknots (*FA* §14)

will be subject to the axioms of geometry.

Frege's understanding of the source of our knowledge of the truths of Euclidean geometry is, he believes, entirely in accord with the Kantian view. But Frege thinks that other aspects of the Kantian view are not quite right. One problem has to do with the importance of different levels of generality. There is, on Frege's view, a limit to the generality of the consequences we can draw from our observations. The general truths that we can establish via evidence of the senses, the synthetic a posteriori truths, can tell us only about the physical, spatio-temporal world. Physical laws are violated in stories in which the drowning haul themselves up out of swamps by their own topknots. The general truths of Euclidean geometry (that is, the synthetic a priori laws), 'govern all that is spatially intuitable, whether actual or product of our fancy' (*FA* §14). Synthetic a priori laws hold over a wider, more general domain than the (synthetic a posteriori) laws of physical science.

But even the laws of Euclidean geometry do not hold everywhere. They do not, in particular, hold in non-Euclidean geometry—the field in which mathematicians draw consequences from sets of axioms that include the denial of one of the axioms of Euclidean geometry. However, there are, Frege claims, laws that do hold in 'the widest domain of all . . . everything thinkable' (*FA* §14). Such laws, for example, as that every object is identical to itself, hold no matter what the subject matter. These laws, Frege thinks, must be laws of logic. Indeed, it is not possible to draw consequences from the denial of a law of logic for, if we try, 'even to think at all seems no longer possible' (*FA* §14). Truths whose justification is purely of a general logical nature must be a priori. But they cannot be synthetic, for synthetic truths hold only of what is spatially intuitable. Only one classification remains: the laws of logic and the truths that follow from them must be analytic.

We can now see the difficulty. An analytic truth, on Kant's characterization, is a truth whose predicate concept is contained in its subject concept. The example we discussed earlier, that all grandmothers are grandparents, fits Kant's characterization. It is easy to identify both predicate and subject concepts and the predicate concept (*grandparent*) is contained in the subject concept (*grandmother*). But it is not obvious that all truths that satisfy Frege's generality criterion of analyticity fit this characterization. For example, the statement that either it is raining or it is not raining appears to be a truth that is a consequence of the general principle that, for any statement on any topic, say *P*, either *P* or not-*P*. Using Frege's generality criterion, this is an analytic truth. Yet this truth does not fit Kant's characterization of analytic truths—it is not obvious that it *has* a subject concept or predicate concept.

Thus Frege offers a new, official, characterization of analyticity: an analytic truth is a truth that can be established by a derivation that relies only on definitions and general logical laws. This new characterization is not meant as a repudiation of Kant's analytic/synthetic distinction. Frege claims, at one point, to be saying what Kant really meant by analyticity. In order to understand the relation between Frege's new characterization of analyticity and the Kantian notion, we will have to look more closely at what logic is.

We often make inferences; that is, we make judgements on the basis of other judgements. Suppose, for example, you know that all whales are mammals and all mammals are vertebrates. You are entitled to the further judgement that all whales are vertebrates. This inference might be represented as follows:

All whales are mammals.
All mammals are vertebrates.
Therefore, all whales are vertebrates.

It should be evident that this inference is a good one. It

simply cannot happen that, although both of the premisses (the first two sentences above) are true, the conclusion (the final sentence) is false. Moreover, there are no gaps in the inference. The truth of the premisses alone guarantees the truth of the conclusion. There is no need to appeal to anything else—to evidence of sense experience, intuition, or to any further facts. We do not even need to know anything about the concept *whale*. For it should be just as evident that the following is a good argument:

All nudibranchs are mammals.
All mammals are vertebrates.
Therefore, all nudibranchs are vertebrates.

Whatever nudibranchs are, provided they are mammals and all mammals are vertebrates, nudibranchs are vertebrates. An inference, or argument, of this sort is said to be *valid*; its conclusion is said to *follow from* its premisses.

Of course, as divers and students of marine biology know, nudibranchs are not mammals, nor are they vertebrates. Both the conclusion and one of the premisses are false. But this does not invalidate the argument. The recognition that an argument is valid is not supposed to provide unconditional licence to judge that the conclusion is true (or that the premisses and conclusion are true). Rather, it provides a conditional licence: a licence to accept the conclusion, provided the premisses are true. Because nudibranchs are not mammals, the validity of the above argument gives us no licence to judge that nudibranchs are vertebrates.

The task of logic is to provide a means for identifying valid arguments. How is this task to be carried out? Let us consider our argument again. The recognition that the argument remains valid when 'nudibranchs' is substituted for 'whales' requires no knowledge about nudibranchs. Indeed, we can substitute any concept-expression for 'whales' in the above argument and the result will still be a valid argument. That is, whatever we substitute for '*A*' in the following, the result will be a valid argument:

All *A*s are mammals.
All mammals are vertebrates.
Therefore, all *A*s are vertebrates.

Nor does our ability to see that this is a valid argument require any knowledge about what is to be a mammal or vertebrate. All we need to see, in order to tell that our argument is valid, is that it has the following form:

All *A*s are *B*s
All *B*s are *C*s
Therefore, all *A*s are *C*s

Arguments that have this form—that is, arguments that we can express by substituting appropriate expressions for '*A*', '*B*', and '*C*' in the above—are said to be *instances* of the argument form.

The strategy of determining whether or not an argument is valid by identifying its form comes from Aristotelian logic. In Kant's time, and in Frege's, Aristotelian logic was generally accepted as providing a means for evaluating arguments whose validity depends on relationships between concepts. The representation of the form of the argument involves a regimentation of each sentence into a subject term and a predicate term. For example, the subject term of the sentence, 'All whales are mammals' is 'whales' and the predicate term is 'mammals'. Aristotelian logic is concerned with syllogisms: arguments in which the predicate of the conclusion appears in one of the premisses and the subject of the conclusion appears in the other premiss. It is not difficult to see the advantage of studying syllogisms. There is a limited number of forms syllogisms can take. The identification of the relatively few valid forms enables one to evaluate a large number of actual arguments.

But what has logic, as it has just been described, to do with analyticity, in Kant's sense? As we saw earlier, on the Kantian view our knowledge of analytic truths is supposed to come from analysis of concepts. It is not difficult to see the following argument as an articulation of a kind of analysis of the concept *grandmother*:

All grandmothers are female grandparents.
All female grandparents are grandparents.
Therefore, all grandmothers are grandparents.

This analysis, of course, consists of a valid argument whose premisses are both analytic truths. The first premiss seems to be an articulation of the content of *grandmother* and, in the second, the predicate concept is clearly contained in the subject concept. On the Kantian view, as Frege understands it, valid derivations whose only premisses are analytic truths should be regarded as a sort of analysis of concepts. Conclusions of such arguments should, therefore, be regarded as analytic. Valid inference is closely connected with Kantian, as well as with Fregean, analyticity.

What of the Kantian assessment of the truths of arithmetic? Kant claimed that, while the proposition 7+5=12 seems, at first, to be an analytic judgement, it is not: an analysis of concepts is not sufficient for us to discover that 12 is the sum of 7 and 5. This is in line with the view that Aristotelian logic can give us all results of conceptual analysis. For we cannot get an Aristotelian proof that 7+5=12 from definitions of the concepts involved. In fact, the connection of Aristotelian logic and analysis of concepts has a great deal of significance for our understanding of the nature of analytic truth. Although Aristotelian logic allows us to recognize a large number of arguments as valid, none of its results are surprising. Thus the proofs of Aristotelian logic cannot give us substantive new knowledge. If the analytic truths are those we can get using only definitions and Aristotelian logic, then their discovery will result in no new knowledge. Kant claimed that the discovery of an analytic truth does not constitute substantive knowledge. Thus he counted mathematical judgements, which clearly do extend our knowledge, as synthetic. Moreover since, on this view, the only substantive knowledge is knowledge of synthetic truths, the only sources of knowledge there can be are pure intuition and sense experience—the sources of our knowledge of synthetic truths.

Although Frege agrees with Kant's assessment of our knowledge of geometry, he parts company with Kant when it comes to the assessment of our knowledge of other mathematical truths. Frege claims that not only the truths of arithmetic, but all truths of mathematics (with the exception of those of Euclidean geometry) are analytic. Given what we have seen so far, this may seem mysterious. A truth is analytic, on Frege's official characterization, just in case it is derivable using only definitions and logic. Although Frege has offered another criterion of analyticity—that truths governing everything thinkable are analytic—it is difficult to see how this criterion accords with his official definition. Moreover, even if we can connect his official characterization of analyticity with this criterion, it is still not obvious why Frege thinks that truths of mathematics other than those of arithmetic are analytic. For he does not actually say why truths of other fields of mathematics govern everything thinkable. To understand Frege's departure from Kant, it will help to know a bit about the history of mathematics.

One of the central fields of mathematics is analysis: the study of infinite processes. Analysis originated in the seventeenth century as a response to the needs of physics and astronomy. Its concerns were understood, originally, to be with continuous magnitudes, such as lengths, areas, velocities, and accelerations—these could be, and often were, represented geometrically. In Kant's time most proofs of analysis exploited techniques of geometry. As long as important theorems of analysis are thought to depend on geometrical proofs, it is reasonable to regard them as synthetic a priori—for synthetic a priori truths are truths whose justification requires appeals to spatial intuition.

In the mid-nineteenth century, however, it became apparent that many of the geometrical proofs were not as secure as they seemed. Some apparently good proofs were identified as fallacious. The difficulties were attributable, in part, to confusions about some of the basic notions of analysis, including those of limit and continuity. The attempt to clarify these notions involved arithmetizing analysis, that is, showing that its truths could be proved from truths of arithmetic. By the time Frege began his work, most proofs of analysis had been separated from geometry and the notion of magnitude. It is not surprising, then, that it would have seemed less evident to Frege that the truths of analysis are synthetic a priori.

But this is still not sufficient to explain the strength of Frege's conviction that these truths are not synthetic a priori. The arithmetization of analysis can show only that in order to determine the source of our knowledge of the truths of analysis (and any other arithmetized field of mathematics) we need to identify the source of our knowledge of the truths of arithmetic. The truths of arithmetic might still be synthetic a priori. Indeed, in *The Elements* Euclid provided a means for showing how to derive truths of arithmetic from truths of geometry. Euclid regarded numbers as magnitudes and represented each whole number as a line segment. This allowed him to present geometrical statements and proofs of fundamental laws of arithmetic. If proofs of this sort are required to justify the truths of arithmetic, then the success of the arithmetization of analysis shows, not that arithmetic rather than geometry is fundamental, but only that mathematicians had misunderstood the role of geometry in the foundations of analysis. It shows, one might argue, that geometry simply needs to come in at an earlier stage—with the justification of the truths of arithmetic. Frege was aware of the existence of this sort of proof of truths of arithmetic and he was also aware of various strategies for defining numbers in terms of geometrical notions. Yet he was still convinced that the truths of arithmetic were analytic. In order to understand why, we need to return to his views about the generality of arithmetic.

The truths of arithmetic, Frege claims, apply to a realm wider than the spatial. Not only spatial objects, but odours, sounds, methods, and ideas can be counted. He infers that, even if it seems that one could rewrite the truths of arithmetic in the language of geometry and prove them from the axioms of geometry, all these proofs would show is that arithmetic holds in the spatial realm. Conclusions of these proofs can only be restrictions of the propositions of arithmetic to the spatial realm. If truths about numbers apply, as Frege claims, to everything, even to such non-spatial entities as methods and ideas, then a more general proof is required. Frege asks, '[H]ow could intuition guarantee propositions which hold for all such heterogeneous quantities, some species of which may still be unknown to us?' (*MC* 1). Only if, Frege believes, the truths of arithmetic apply to the widest domain of all; to everything thinkable.

This marks a departure from Kant. On Kant's view reason alone cannot give us substantive knowledge—all substantive knowledge must have pure intuition or sense experience as a source. On Frege's view, however, our knowledge of the most general truths is substantive but does not have either sense experience or pure intuition as its source. Thus Frege regards reason itself as a source of knowledge. Just as intuition underlies the possibility of perception, reason underlies the possibility of thought. Frege's aim is to show that reason alone is the source of our knowledge of the truths of arithmetic; that the truths of arithmetic can be derived using solely logical means.

We have already seen one of the obstacles that must be surmounted. Aristotelian logic, the logic already available to Frege, is not sufficiently powerful to provide proofs of the truths of arithmetic. Thus part of Frege's project is to replace Aristotelian logic with a new, more powerful logic. How is this to be done? One might suspect that the first step is simply to add the basic inferences of arithmetic to the valid inferences of Aristotelian logic. Yet this strategy will not do. In order to show that arithmetic is analytic, Frege must show that the only principles needed to infer the conclusions of these inferences from their premisses are principles that hold, not just of the domain of numbers, but of the widest domain of all. This task is not at all trivial. For the reasoning and principles involved in many proofs of truths of arithmetic do not appear to be applicable in the widest domain of all. One of the features that seems to distinguish reasoning about natural numbers (i.e., 1, 2, 3, etc.) is the use of a particular principle that is peculiar to this domain: mathematical induction.

What is mathematical induction? Suppose we want to show that all natural numbers have a particular property. According to the principle of mathematical induction, we need to establish only two things: first, we need to establish that the property holds of 1. And second, we need to establish that for any number, *n*, if the property holds of *n*, it holds of the successor of *n* (if this is so, the property is said to be 'hereditary in the natural number sequence'). To see that this suffices to show that the property in question holds of all natural numbers, let us suppose it has been shown that the property satisfies both of the above conditions. First, the property certainly holds of 1, since this has been shown explicitly. It has also been shown that the property holds of the successor of *n*, provided it holds of *n*. Thus, since the property holds of 1, it holds of the successor of 1, that is 2. Since the property holds of 2, it holds of the successor of 2, that is, it holds of 3. It should be evident, from this, that the principle of mathematical induction does not lead us astray. But it also seems that the success of the principle of mathematical induction is based on the nature of the domain of natural numbers. It works because the sequence of natural numbers consists of 1, the successor of 1, the successor of the successor of 1, etc. But if mathematical induction is a form of inference peculiar to the domain of numbers, its results will not be derivable from logic alone.

One of Frege's insights is that the principle of mathematical induction is not a special principle derived from the peculiar nature of the domain of numbers. In particular, the feature of the natural number sequence that allows us to use mathematical induction to prove general truths about numbers is not peculiar to the natural number sequence. Consider the relation that holds between two people, *x* and *y*, when *x* is a direct ancestor of *y*—i.e. when *y* is a child of *x*; or *y* is a child of a child of *x*; or *y* is a child of a child of a child of *x*, etc. The sequence of natural numbers (i.e. 1, the successor of 1, the successor of the successor of 1 . . .) is very like a sequence consisting of a particular person and her direct descendants (e.g. Anna, Anna's children, Anna's children's children, etc.). Both sequences are determined by a first member and relation—in one case, by the number one and the *is the immediate successor of* relation; in the other by Anna and the *is a child of* relation.

Now suppose we wish to argue that a particular property, for example the property of having brown eyes, holds of all members of the sequence consisting of Anna and her direct descendants. It is possible to use an argument almost indistinguishable from mathematical induction. The principle of mathematical induction tells us that, in order to show that all members of the natural number sequence have some particular property we need only establish that the first member has the property and that the property is hereditary in the natural number sequence. Let us suppose that we have established the corresponding facts for the sequence consisting of Anna and her descendants. That is, suppose it has been established that the first member of the sequence (Anna) has brown eyes and suppose it has been established that the property of having brown eyes is hereditary—that is, every child of a brown-eyed person has brown eyes. We can infer that every member in our sequence has brown eyes. If there is any doubt, we can go through the same reasoning that was used to convince us that mathematical induction proofs work in arithmetic.

Of course, the conclusion of the latter argument is synthetic a posteriori—it can only be established by making appeals to evidence of the senses. But the only role these appeals play is in establishing the truth of the premisses. As we saw earlier, a valid argument need not have true premisses. To determine that this is a valid argument is to determine that the truth of the premisses is all we need to license our judgement that the conclusion is true. Using this standard, the above argument seems to be valid. We can see, without any appeals to the senses, intuition or any further facts that, provided the premisses of this argument are true, the conclusion is true. Indeed, all that seems to be required is an analysis of the concepts and relations involved.

These observations provide a reason for thinking that inferences exploiting mathematical induction do not rest on features peculiar to the domain of numbers. Mathematical induction, as we saw earlier, licenses us to infer that, provided a property holds of the first member of the natural number sequence and is hereditary in the natural number sequence, it holds of every member of the natural number sequence. But, as we have just seen, the legitimacy of this sort of inference does not rest on any peculiar feature of the numbers. We can similarly infer that, provided a property holds of the first member of the sequence consisting of Anna and her descendants and provided it is hereditary in this sequence, the property holds of every member of this sequence. Nor does the legitimacy of either inference rest on features peculiar to people and numbers. All that seems to matter is the structure of the sequence—in particular, the fact that a first member and a relation determine each sequence. And there is no reason to think that there can be sequences only of certain sorts of objects. Inferences about sequences and their properties can be inferences about sequences of any sort of objects. That is, these inferences have the hallmarks of inferences justified by logic alone.

Aristotelian logic tells us otherwise. For Aristotelian logic depends on an understanding of judgements that makes it impossible to recognize complex arguments involving relations as logically valid. Frege's first task was to substantiate his conviction that these general inferences about sequences are valid inferences. He did this by rejecting the subject/predicate regimentation on which Aristotelian logic depends and developing a new, more powerful logic. The result, his revolutionary new logic, was introduced in his 1879 monograph, *Begriffsschrift.*

# 3 *Frege's new logic*

In order to understand the revolutionary nature of Frege's new logical regimentation, it will help to begin by looking at some of the limitations of the logical methods available before his work. Frege was not the only one to recognize that Aristotelian regimentation cannot always be used to evaluate inferences correctly. It was well known long before Frege undertook his project that some logically valid inferences are not valid Aristotelian syllogisms. The following inference is an example:

If Ralph is a parrot, then Ralph is not a nudibranch.
Ralph is a parrot.
Therefore, Ralph is not a nudibranch.

It is obvious that this is a valid argument. We may need evidence of the senses to determine whether the premisses are true or not. However, no evidence from the senses or intuition—and no evidence other than what is already available in the inference—is required to see that the conclusion follows from the premisses. That is, the inference seems to be logically valid. But this inference is not a syllogism at all. A syllogism is an argument in which the predicate of the conclusion appears in one of the premisses and the subject of the conclusion appears in the other premiss. And the first premiss of the above argument does not even have subject/predicate form. Rather, this premiss is a complex statement, called a conditional, which is formed by combining two statements: an antecedent,

Ralph is a parrot

and consequent

Ralph is not a nudibranch.

Moreover, although the second premiss does have subject/predicate form, we do not need to know that it has this form (or what its subject and predicate are) in order to evaluate the inference. If our interest is in evaluating this inference, then the constituents of the premisses and conclusion that must be represented are statements, not subjects and predicates. What is important is that one of the premisses is a conditional; the other premiss is the antecedent of the conditional; and the conclusion of the inference is the consequent of the conditional. This is a long way of saying that the inference has the following form:

If *B* then *A*
*B*
Therefore, *A*

All letters that appear in this argument form are place-holders, not for concept expressions (as they were in the arguments about mammals and vertebrates that we considered in the previous chapter), but for sentences. The formal part of the argument seems to consist of relationships between propositions, rather than concepts. A logic that evaluates these sorts of arguments is called a propositional logic. This particular sort of inference is said to be licensed by the rule *Modus ponens.*

Although propositional logic and Aristotelian logic were regarded as competitors in the early history of logic, by the nineteenth century both were accepted as important contributions to the general project of evaluating inference. But they were also considered distinct. It was a common view that Aristotelian logic provided a method for evaluating inferences whose validity was based on relations between concepts, while propositional logic provided a method for evaluating inferences based on relations between propositions. The two methods exploit distinct ways of viewing and representing the construction of the premisses and conclusion of an inference.

Is there any way to construct a unified logic that makes it possible to recognize valid inferences of both sorts? One of

Frege's most illustrious predecessors, George Boole (1815–64), introduced a new symbolic notation in which it was possible to represent both syllogisms and propositional arguments. Once an inference is expressed in Boole's notation, one can go on to evaluate the inference by performing a mechanical calculation. But this technique still involves a sharp separation between the two sorts of inference. For while Boole's notation can represent the logical structure that has significance either for propositional or for Aristotelian inference, it cannot represent both at once. This is a result of Boole's use of symbols. A symbol from Boole's notation, when it is used in the expression of a syllogism, has a meaning that is different in kind from the meaning it has when it is used in the expression of a propositional argument. In order to use Boole's technique to evaluate an argument, one must decide first what sort of argument it is and represent its form accordingly. One consequence is that, for many statements, Boole's notation cannot be used to represent all content that has significance for inference. Consider, for example, one of the premisses of the above argument:

If Ralph is a parrot, then Ralph is not a nudibranch.

This statement has both propositional complexity (it is a conditional) and subject/predicate complexity (its antecedent and consequent both have subject/predicate form). Whenever this statement is represented in Boole's notation some of this complexity will be omitted—as it is in the above representation of the form of the inference licensed by *Modus ponens*.

Given a statement, Boole's notation cannot be used to express all content that has significance for any inference in which it can figure. But is this a defect of Boole's notation? After all, if we want a technique for evaluating particular inferences, it should suffice that we can express, for any inference, all content of its premisses and conclusion that are significant for the evaluation of this particular inference. Boole's technique enables us to identify as valid all arguments that can be identified as valid either by Aristotelian or by propositional logic. If these are, indeed, the only valid arguments, then Boole's notation is sufficient for purposes of evaluating arguments. However, should it turn out that there are valid arguments whose validity is due to a combination of propositional and non-propositional complexity, Boole's method will be deficient. Anyone who thinks that there are valid arguments of this sort will want a notation with more expressive power than Boole's.

It was Frege's conviction that there are valid arguments that cannot be identified either as valid Aristotelian or as valid propositional arguments. He set out to construct a new logical notation, Begriffsschrift (or, concept-script); a notation designed to express, for any statement, *all* content that has significance for *any* of the inferences in which it can figure. He called this the 'conceptual content' of a statement. It is easy to see the difference between a notation that expresses conceptual content, in this sense, and Boole's notation. An expression of a statement in Boole's notation represents only a part of its content—the part that has relevance for the evaluation of a particular inference. In order to know how a statement should be represented in Boole's notation, one must know what inference is under consideration. In contrast, the expression of a statement in Begriffsschrift is independent of any one inference in which it can appear. For the rendering of a statement in Begriffsschrift must express all content that is relevant to *any* inference in which it can appear. Frege wanted to develop a notation that can be regarded, not simply as a tool to be used in determining the validity of arguments, but as a language.

Although Begriffsschrift is meant to be a language, there are important differences between Begriffsschrift and natural language—the sort of language we use in everyday life. Frege's language is artificial and it is not designed to be used for everyday purposes. As he mentions in the preface to *Begriffsschrift*, when he began his project he found that natural language was not adequate for his purposes. In natural language it is difficult to express complex conceptual contents precisely. Also, as Frege argues throughout his career, natural language has a number of logical defects. His aim, however, is not to improve natural language or to replace it with a logically perfect language. Rather, he regards his logically perfect language, Begriffsschrift, as a scientific tool. Frege's notation is designed, as Boole's was, to be used in connection with a method for evaluating inferences.

In the preface to *Begriffsschrift* Frege compares his logical language to the microscope. His logical language is useful, just as the microscope is, for certain scientific purposes and entirely useless for others. For most purposes, we are better off using the naked eye to make our observations than we are using a microscope. Similarly, for most purposes, we are better off expressing our statements in natural language than we are expressing them in Begriffsschrift. It is *only* when our aim is to evaluate inferences that Begriffsschrift is a better instrument for expressing our statements than natural language. The aim is to introduce a system of evaluation that will make it a mechanical task, once an inference is expressed in Begriffsschrift, to determine whether or not the inference is correct and gapless. This involves the introduction of a small group of primitive logical truths as well as a rule that licenses inferences of a particular form.

We will turn, shortly, to the details of Frege's new logic and the sense in which it is an improvement over earlier logics. Before we do this, however, it will help to say a bit more about words and their contents and to introduce explicitly some terminology that I have already used extensively.

Any discussion of conceptual content requires the recognition of a distinction between words and their content. This can be illustrated by considering words 'and' and 'but'. Let us consider what we are entitled to infer from the following true claim about an ostrich:

(*) It is a bird, but it cannot fly.

We are entitled to infer that it is a bird, for the truth of (*) *guarantees* that it is a bird. If the ostrich were not a bird, then (*) would not be true. Similarly, we are entitled to infer that it cannot fly. It may seem that we are also entitled to infer something more: that most birds can fly. But this is a mistake. For the truth of (*) does not guarantee that most birds can fly. If most birds could not fly, it would *still* be true that the ostrich is a bird, but it cannot fly—although it might be odd to use the word 'but'. On Frege's view, we are entitled to infer, from

*A* but *B*

exactly the same statements that we are entitled to infer from

*A* and *B*.

That is, 'and' and 'but' have the same significance for inference; they have, to use Frege's technical term, the same conceptual content. Thus Frege's project requires us to distinguish words from the content associated with them. Of course, this is not to say that we associate exactly the same content with 'and' and 'but'—if we did, it would not be odd to use 'but' in the circumstance described above. Frege's project also requires us to recognize a particular sort of content—conceptual content—that Begriffsschrift is designed to express.

We can find, not only pairs of distinct words, but also pairs of distinct sentences that have the same conceptual content. Frege uses the following pair of sentences as an example:

Hydrogen is lighter than carbon dioxide.

And

Carbon dioxide is heavier than hydrogen.

Both sentences express something that can be either true or false. In *Begriffsschrift,* Frege refers to the conceptual content of—or what is expressed by—this sort of sentence as a 'content that can become a judgement'. I will use this expression as well as the expressions 'proposition' and 'statement' for what is expressed by a sentence that can be either true or false.

We are now in a position to begin examining Frege's logical notation. Begriffsschrift is to be used for the expression of inferences. What will the expression of an inference look like? Inferences are strings of assertions. In an inference some propositions, the premisses, are simply asserted and another proposition is asserted on the basis of the premisses. But not every expression of a proposition is an assertion. Consider the following inference:

If 3 is a square root of 4, then 3 is a cube root of 8.
3 is not a cube root of 8.
Therefore, 3 is not a square root of 4.

The sentence '3 is a square root of 4' appears as part of the first premiss. Thus the proposition this sentence expresses is expressed in the first premiss. But it is not asserted in the first premiss. Indeed, the whole point of expressing it in this inference is to show why it should not be asserted. In the conclusion, it is denied.

The difference between asserting and expressing is indicated in Frege's Begriffsschrift by the first two signs he introduces: the signs he calls the 'judgement stroke' and the 'content stroke'. The judgement stroke is a small vertical line and the content stroke is a long horizontal line. The representation, in Begriffsschrift, of a content that can become a judgement (a propositional content) is always preceded by a content stroke. If this content is asserted in an inference, the content stroke is preceded by the judgement stroke. For example, in the first inference we looked at in this chapter, the inference licensed by *Modus ponens*, one of the premisses was: Ralph is a nudibranch. Supposing, for the moment, that 'Ralph is a nudibranch' is a sentence in Frege's logical language, then this premiss could be represented in the inference as follows:

|—— Ralph is a nudibranch

Begriffsschrift also has two symbols for representing propositional complexity: the condition stroke and the negation stroke. Let us begin with the condition stroke. The second premiss in our *Modus ponens* inference is a conditional: If Ralph is a parrot, then Ralph is not a nudibranch. This is a complex proposition with two propositional constituents. How is it to be represented in a logical notation? One might suppose that Frege's task is to replace the natural language expression 'if

. . . then . . . ' with a notation that represents the content of the natural language expression. But this is not quite how Frege regards his task. For the sort of conditional that, he claims, is expressed by use of the word 'if' is a causal conditional. He gives as an example,

> If the Moon is in quadrature with the sun, the Moon appears as a semicircle.

What is expressed in this statement, he suggests, is a causal connection between the Moon's being in quadrature with the sun and the Moon's appearing as a semicircle. But causal connections occur only in the physical realm. There should not be a *logical* symbol for relationships that hold only in the physical realm. Of course, it is not quite right to say that use of the locution 'if . . . then . . . ' invariably expresses a causal relation. After all, this locution is also used in inferences of geometry and arithmetic, inferences that depend on no causal relations. Frege's aim is to introduce a Begriffsschrift symbol for the sort of conditional that can be used regardless of subject matter.

This is the sort of conditional that is used in the inferences of propositional logic. The following form, the form of inference licensed by *Modus ponens*, is one sort of valid propositional inference:

> If *B* then *A*
> *B*
> Therefore, *A*

If we agree that all inferences of this form are valid, this tells us something about the content of the conditional. The conditional must provide the information needed to allow us to infer the consequent from the antecedent. That is, we cannot be entitled to accept the conditional and its antecedent without also being entitled to accept its consequent. The minimal content required of a conditional, then, is that it rule out the possibility that the antecedent is true but the consequent is false. Moreover, as the widespread acceptance of *Modus ponens* indicates, this content is inherent in our everyday use of the expression 'if ... then ... '. This is the content that Frege's conditional is meant to capture.

How are we to understand this content? Since the conditional is to be regarded as ruling out certain possibilities, it will help to begin by listing all possibilities. There are two propositions that concern us: *B*, the antecedent, and *A*, the consequent. As a result we have four possibilities, which Frege lists as follows:

I *A* and *B*
II *A* and not-*B*
III not-*A* and *B*
IV not-*A* and not-*B*

The conditional (if *B* then *A*), he says, denies the third of these possibilities. One notable feature of this sort of conditional is that its truth is entirely determined by whether the antecedent and consequent are true or false. Such a conditional is called a 'truth-functional' conditional. Frege illustrates this with the following example:

If the sun is shining, then $3 \times 7 = 21$.

Since the consequent is true, we can rule out the third possibility, the possibility that the antecedent is true and consequent false. But there is no causal connection, indeed no connection at all, between the sun's shining and its being true that $3 \times 7 = 21$.

But has Frege chosen the correct truth-functional conditional for his purpose? Although it is clear that the conditional must rule out the possibility that the antecedent is true and the consequent false, one might suspect that the conditional should rule out other possibilities as well. Why should all conditionals with false antecedents be true? Suppose that Ralph *is not* a parrot. It may not seem that in this situation we really think it is true that *if* Ralph is a parrot, *then* he is a nudibranch.

It is important to notice, first, that taking the conditional to be true in this situation will not get us into any trouble in our inferences—since Ralph is not a parrot, we cannot infer anything else about Ralph from the conditional. But this is not enough to explain why we *should* take the conditional to be true. In order to understand why we should take conditionals with false antecedents to be true, we need to look more closely at some of the work Frege's logic needs to do.

This is easiest to see if we remember that Frege wants to find a method for looking at both propositional and non-propositional conceptual content simultaneously. Consider the true claim that every cube root of 8 is a square root of 4. On Frege's analysis, this tells us that,

> for anything we choose, say *n*, if *n* is a cube root of 8, then *n* is a square root of 4.

Given this analysis—and given that it is true that every cube root of 8 is a square root of 4—then, no matter what *n* is, it must be true that if *n* is a cube root of 8, *n* is a square root of 4. That is, all the following must be true:

(*a*) If 1 is a cube root of 8, then 1 is a square root of 4.
(*b*) If 2 is a cube root of 8, then 2 is a square root of 4.
(*c*) If –2 is a cube root of 8, then –2 is a square root of 4.

We can now show why every truth-functional conditional with false antecedent must be true. The truth of a truth-functional conditional must be entirely determined by whether the antecedent and consequent are true or false. Consider (*a*). It is a conditional whose antecedent and consequent are both false. Since (*a*) is true, *every* conditional whose antecedent and consequent are both false must be true. Now consider (*c*). It is a conditional whose antecedent is false and whose consequent is true. Since it is true, *every* conditional whose antecedent is false and whose consequent is true must be true.

Because Frege's analysis requires that both (*a*) and (*c*) be true—and because his logic requires that the truth of a conditional be entirely determined by the truth or falsehood of antecedent and consequent—every conditional with false antecedent must be true. Moreover, since (*b*) has true antecedent and true consequent, every conditional whose antecedent and consequent are both true must be true. That is, the conditional must be understood as excluding *only* one of the four possibilities: the case in which the antecedent is true and the consequent false.

In Begriffsschrift, conditionals are represented by preceding both the antecedent-expression and consequent-expression with content strokes and connecting the content strokes with a vertical line that Frege calls a 'condition stroke'. Because Frege's actual symbols are difficult to print and to read, from this point on I will use contemporary symbols. One contemporary symbol for the conditional is an arrow. We can write, for example:

1 is a cube root of 8 → 1 is a square root of 4.

After introducing the condition stroke, Frege informally introduces *Modus ponens* as a rule of inference by noting that his explanation makes it apparent that from two judgements $B$ and $B \rightarrow A$, the new judgement $A$ follows.

Frege's other symbol for propositional complexity is his negation stroke. The negation stroke is used to express the denial of a proposition. I will be using the symbol '~' instead of Frege's negation stroke. Supposing '–2 is a cube root of 8' were a Begriffsschrift expression, we would represent the claim that –2 is *not* a cube root of 8 as follows:

~(–2 is a cube root of 8).

Conditional and denial are the only propositional connectives for which Frege has symbols. Most systems of propositional logic used today include symbols for other propositional connectives as well. For example, in most propositional systems there is a special symbol to express the conjunction of two propositions ($A$ and $B$). However, Frege's two symbols have enough power to express all propositional complexity. For example, the conjunction of $A$ and $B$ can be expressed using only the arrow and '~' as follows:

$\sim(A \rightarrow \sim B)$.

To see this, remember that the conditional rules out exactly one possibility—that the antecedent is true and the consequent false. Thus

$A \rightarrow \sim B$

rules out the possibility that $A$ is true and $\sim B$ false or, in other words, that $A$ is true and $B$ is true. Hence the denial of this conditional is true, just in case $A$ and $B$ are both true. Although Frege limits himself to symbols for conditional and denial, I will sometimes depart from this, for the purpose of convenience, and use the contemporary symbol, '&', for conjunction.

The next symbol Frege introduces is his sign for identity, a triple bar. Although I will be using '=' instead of Frege's triple bar, it is important to realize that Frege's notion of identity is not a familiar one. Let us begin with the familiar notion. In arithmetic we put the identity sign between number names. The number names can be simple numerals (e.g., '1', '2', etc.) or they can be complex (e.g. '2×(3+5)'). Consider the following statement:

$2\times(3+5)=4\times4.$

What does this tell us? Frege's answer, in *Begriffsschrift*, is that it tells us that the two signs '2×(3+5)' and '4×4' have the same content. Although one might suspect that it would be immediately apparent when two signs have the same content, a moment's reflection should show that when computations become sufficiently complex, it will be far from clear whether or not two signs are signs for the same number. With each sign of this sort is associated a different way of determining which number it picks out. And it can be informative to discover that two ways of determining a number give us the same result. Frege describes the content of '$A=B$', as follows

The sign $A$ and the sign $B$ have the same conceptual content, so that we can everywhere put $B$ for $A$ and conversely. (*B* §8)

It may not seem immediately evident that there is any difference between Frege's use of the identity sign and ours. But now consider, again, Frege's claim that his identity symbol expresses identity of content between signs. Although Frege distinguishes sentential expressions from non-sentential expressions (in particular, the content stroke can only be prefixed to a sentential expression), all Begriffsschrift expressions have conceptual content. Thus Frege's identity sign can appear, not only between number names and similar expressions, but also between sentential expressions. Supposing that 'Hydrogen is lighter than carbon dioxide' and 'Carbon dioxide is heavier than hydrogen' were Begriffsschrift sentences, the following would be a Begriffsschrift sentence:

> (Hydrogen is lighter than carbon dioxide)=(carbon dioxide is heavier than hydrogen).

Although Frege's view of the identity sign changed over the course of his career, he continued to use the identity sign as something that could appear, not only between two names, but also between two sentential expressions. To see why this is so, we need to turn, finally, to one of Frege's central departures from Aristotelian logic: his rejection of the use of subject/predicate regimentation as a basis for the expression of the conceptual content of propositions.

Frege's Begriffsschrift regimentation is based on a view of the simplest sort of statement as having, not subject/predicate form, but function/argument form. These notions of function and argument have their origins in mathematics. In order to understand this, we will need to look at these mathematical notions. To begin, what is a function? One answer is that a function is some sort of operation or process of transformation. Given an object, perhaps of some particular kind (this object is called an 'argument') a function returns a value—another, or perhaps the same, object. We can think of the immediate successor of a particular number as the value of a function—the immediate-successor function—on that number. Given the number 1 as argument, the immediate-successor function returns, as value, the number 2; given 2 as argument, the immediate-successor function returns, as value, the number 3, etc. Since each natural number has a unique immediate successor, this function can take any natural number as argument.

The immediate-successor function is called a 'one-place' function because it takes only one object as argument. Other familiar functions take more arguments. The addition function takes two numbers as arguments and returns, as value, their sum. For example, given the numbers 1 and 2 as arguments, the addition function returns, as value, the number 3. We can think of a function as something that has blanks (argument spaces) that can be filled in by any of a large number of arguments. Only when the blanks of the function are filled by arguments do we get a value.

In mathematical contexts, we make general claims about functions by using complex symbols that indicate where the argument places are. We use the expressions '$x+y$' for the addition function and '$x+1$' for the immediate-successor function. The letters in these expressions are called 'variables' and they indicate where the arguments are to go. When in the expression '$x+y$', '$x$' and '$y$' are replaced with number names, the resulting expression is a name of the value the addition function has on the numbers named. '$2+3$' is a name for a number: the value the addition function returns when the function is applied to the arguments 2 and 3.

How do the mathematical notions of function and argument form the basis of a logical language? Let us return to one of the sentences we looked at earlier:

Hydrogen is lighter than carbon dioxide.

From the point of view of Aristotelian logic, and from the point of view of the grammar of natural language, this sentence should be understood as having subject/predicate form. Its subject is 'hydrogen' and its predicate is 'is lighter than carbon dioxide'. It is worth noting, however, that there is something fishy about this Aristotelian analysis. We think of the subject of the sentence as what the sentence is about. Yet, while the subject of the sentence is 'hydrogen', it is no less about carbon dioxide than it is about hydrogen. In contrast, on Frege's analysis, there is no subject or predicate.

Instead of looking at the grammatical structure of each sentence, Frege asks the reader to consider replacing the word 'hydrogen' in the first sentence with various other words, for example, 'oxygen' and 'nitrogen'. He asks us, that is, to consider hydrogen, oxygen, and nitrogen as arguments of the function: ____ *is lighter than carbon dioxide.* The values of the function for these arguments are the conceptual contents of the claims, respectively:

> Hydrogen is lighter than carbon dioxide,
> Oxygen is lighter than carbon dioxide.

and

> Nitrogen is lighter than carbon dioxide.

As we have seen, Frege notes that the sentence

> Carbon dioxide is heavier the hydrogen.

expresses the same statement that our original sentence expresses. One might be inclined to analyse this statement as the value of the function: ____ *is heavier than hydrogen* for carbon dioxide as the argument.

So far, it may seem that there is no important difference between Frege's regimentation and the Aristotelian regimentation. After all, each of the above sentences contains a perfectly good Aristotelian predicate, either '____ is heavier than hydrogen' or '____ is lighter than carbon dioxide'. Each contains a perfectly good subject 'oxygen', 'hydrogen', 'nitrogen', or 'carbon dioxide'. But, on the function/argument view of statements, there is yet another strategy for looking at this statement—the strategy that avoids the choice forced upon us by Aristotelian regimentation. We can view both carbon dioxide and hydrogen as arguments of a two-place function: . . . *is lighter than* ____. On this regimentation there is no subject—the statement is as much about carbon dioxide as it is about hydrogen. Nor is there a predicate—instead of a predicate we have a function that gives us a complete statement when it is completed by putting arguments in its two blank spaces. This function can also be viewed as a two-place relation: a relation that holds between two objects when the first is lighter than the second.

This regimentation seems to express more of the complexity of the statement in question than the Aristotelian regimentations. In particular, the introduction of two-place functions as constituents of conceptual contents allows Frege to express the sort of complexity that is needed if he is to be able to show that general truths about sequences can be derived using logic alone. For, as we are now in position to see, the general truths about sequences are dependent on characteristics of two-place relations.

We saw earlier that Frege's conviction that the truths of arithmetic are analytic comes, in part, from his conviction that mathematical induction is an application of a general truth about sequences. The number sequence consists of 1, the immediate successor of 1 (i.e. 2), the immediate successor of 2 (i.e. 3), etc. One of the fundamental features of this sequence is that it is ordered by a two-place relation, the relation that holds between two numbers when the second is a (perhaps not immediate) successor of the first. Or, to use a more familiar locution, the sequence is ordered by the less-than relation. The number 1 is the least in the sequence: 1 is less than 2, 2 is less than 3, etc. One way of describing the structure of the sequence is to describe the less-than relation that orders this sequence. There is a least member of the sequence (a member that is less than every other member), and for every member of the sequence there is a greater member of the sequence. Progress up the sequence results in increasing size. That is, for any three members of the sequence, say, $x$, $y$, and $z$, if $x$ is less than $y$ and $y$ is less than $z$, then $x$ is less than $z$. These features of the number sequence, as ordered by the less-than relation, can be expressed using Begriffsschrift, but not using Aristotelian regimentation.

Although it would take us too far afield to investigate exactly how Aristotelian regimentation fails, it is not difficult to get a general idea of the problem. Consider one of the features of the less-than relation, the fact that progress up the sequence results in increasing size—that is, for any members of the sequence *x*, *y*, and *z*, if *x* is less than *y* and *y* is less than *z*, then *x* is less than *z*. In the next few paragraphs I will abbreviate this rather long claim by saying that the less-than relation is transitive. Given that the less-than relation is transitive—and given that 1 is less than 2 and 2 is less than 3—we can infer that 1 is less than 3. That is, our logical regimentation ought to be sufficiently powerful to exhibit the features that make the following argument valid:

1 is less than 2
2 is less than 3
The less-than relation is transitive
Therefore, 1 is less than 3

It is obvious that there will be some sort of difficulties involved in representing the form of the third premiss, the claim about transitivity. However, even if we restrict our attention to the first two premisses and the conclusion, Aristotelian regimentation seems unable to represent all the content that has significance for this inference. The first two premisses and the conclusion, it seems, should have subject/predicate form. Indeed, at first glance it may seem evident how to begin our translation. For the conclusion and the second premiss seem to share a predicate (*is less than 3*). In each of these statements something (in one case 2, in the other 1) is said to have this property. Since we want to represent the form of each of these statements, we will replace the shared predicate with a letter. Let us use '*P*' to represent the predicate. Now, how are we to represent first premiss? It would seem, given our earlier decisions, that its subject is 1 and its predicate is *is less than 2*. Since we have no symbol for this predicate, we will introduce a new one: *R*. Our representation, so far, is:

1 is *R*
2 is *P*
The less-than relation is transitive
Therefore, 1 is *P*.

It should be clear that something has gone wrong. For the purpose of logical regimentation is to exhibit those features of the argument that lead us to believe it is valid. What has happened in the above attempt to give a logical regimentation of our argument is that some of the content that is important for its evaluation has disappeared. In the first premiss, for example, the representation of the form obscures the fact that the statement tells us something about 2. Yet the validity of the inference is due, in part, to the fact that both of the first two premisses tell us something about 2.

This could be rectified, of course. If we want to view the first premiss as making a statement about the number 2, we can regard 2 as the subject and *is greater than 1* as the predicate. We need, then, to introduce a new predicate letter, say, '*S*'. Our representation becomes:

2 is *S*
2 is *P*
The less-than relation is transitive
Therefore, 1 is *P*

But this regimentation also will not do. For we have now left out something else of importance for our evaluation of the inference: that the first premiss tells us something about 1. The problem is that our first premiss is about both 1 and 2. Neither of these can be a predicate, but Aristotelian regimentation requires us to pick only one subject—1 and 2 cannot *both* be subjects of the first premiss.

We can now see one of the benefits of Frege's decision to abandon subject/predicate regimentation in favour of function/argument regimentation. The introduction of two-place functions as constituents of conceptual contents allows us to represent the first premiss as having both 1 and 2 as constituents. The constituents of the first premiss, using Frege's regimentation, are 1, 2, and a two-place function: ____ *is less than* . . . . This sort of regimentation allows us to see that the first premiss is about both 1 and 2; that the second premiss is about both 2 and 3 and that the conclusion is about both 1 and 3. It also allows us to see that there is a constituent that is common to all three of these statements: the less-than relation. Using an upper case letter, '*L*', to represent the function, we can now represent the argument as follows:

> $L\,(1, 2)$
> $L\,(2, 3)$
> The less-than relation ($L$) is transitive
> Therefore, $L\,(1, 3)$

What about the final premiss?

The final premiss is an abbreviation of a complex statement and, to make the structure of the complex statement clear, it will help to use the mathematical symbol '$<$' as an abbreviation of 'is less than'. To say that the less-than relation is transitive, then, is to say: for any members of the natural number sequence, say $x$, $y$, and $z$, if $x<y$ and $y<z$ then $x<z$. Can Frege's notation express the conceptual content of this claim? That is, can Frege's notation express all the content of this claim that has significance for inference? Let us consider what we can infer from this premiss. Unlike the other statements, each of which is about a particular pair of numbers, the final premiss is a general statement. It tells us that, no matter what $x$, $y$, and $z$ are, if $x<y$ and $y<z$ then $x<z$. Thus, among the statements that we are allowed to infer from the general statement are:

> If $1<2$ and $2<3$, then $1<3$.
> If $2<1$ and $1<3$ then $2<3$.
> If $4<3$ and $3<1$ then $4<1$.

This list of statements should look familiar, for this is not the only general claim we have examined in this chapter. Another such claim is that every cube root of 8 is a square root of 4, whose consequences include:

If 1 is a cube root of 8, then 1 is a square root of 4.
If 2 is a cube root of 8, then 2 is a square root of 4.
If –2 is a cube root of 8, then –2 is a square root of 4.

The content of this claim is that, whatever we may take for its argument, the function *if x is a cube root of 8 then x is a square root of 4* yields a truth or—to use Frege's locution in *Begriffsschrift*—is a fact.

In order to express generality in Begriffsschrift, Frege introduces quantifier notation. To see how the notation works, let us begin by expressing the general statement in stilted English:

> Choose anything you like, call it '*x*', if *x* is a cube root of 8 then *x* is a square root of 4.

Or:

> For any *x*, if *x* is a cube root of 8 then *x* is a square root of 4.

The expression that begins the sentence ('for any *x*') is a quantifier expression. The places in which '*x*' appears, in the remainder of the sentence, are all places that could be occupied by a number name. But '*x*' is a variable, not a number name. Unlike '1', '*x*' is not used to talk about some particular object but rather, is used in combination with a quantifier to express generality. Our understanding of the occurrences of '*x*' that appear in the sentence are all tied to the quantifier expression that begins the sentence. In this case we say that all the occurrences of '*x*' in this sentence are bound by the quantifier expression, or that the quantifier expression has the entire sentence in its scope. Frege introduces a notation in his logical language to replace the stilted natural language quantifier expression 'for any *x*'. Once again, for purposes of convenience, I will use a contemporary notation, '(x)', rather than Frege's actual notation. Supposing, for the moment, the expressions 'cube root of 8' and 'square root of 4' are Begriffsschrift expressions, the translation of our general claim is:

> (*x*) (*x* is a cube root of 8 → *x* is a square root of 4).

Frege's technique for expressing generality has two important features. One of these is the use of variables. It is easy to see how useful variables are by considering what is involved in stating that the less than relation is transitive. The statement, without using variables, would be something like this:

> for any three things, if the first is less than the second and the second is less than the third, then the first is less than the third.

The claim is much clearer, however, if it is expressed using variables:

> for any $x$, $y$, and $z$, if $x < y$ and $y<z$ then $x<z$.

A second important feature of Frege's technique for expressing generality is his use of quantifiers, which includes an indication of the scope of the generality. Although each of the quantifiers we have seen so far has the entire sentence in its scope, this is not the only way to use quantifiers. In order to be able to express the conceptual content of mathematical inferences, Frege also needs to be able to use quantifiers that have only a part of the sentence in their scope. It is easiest to see the importance of marking quantifier scope by looking at a non-mathematical example. Consider the following two sentences:

> Everything is either green or not green.

And

> Everything is green or everything is not green.

The first of these is, presumably, true while the second is, clearly, false. The difference is clearly indicated by the use of the English quantifier word 'everything'. While in the first case the scope of this word is the entire sentence, in the second there are two quantifier words, each of which has only a part of the sentence as its scope. Frege's quantifier notation allows us to indicate this difference as follows:

> ($x$) ($x$ is green or $x$ is not green)

and

$(x)$ ($x$ is green) or $(x)$ ($x$ is not green).

There is also another sort of quantified statement that we must be able to express in logical notation. This is what is called an 'existential' statement. An example is: there is a number that is $<2$. It is customary, today, to introduce a special quantifier notation for this sort of statement. We would write: $(\exists x)$ (x is a number that is $<2$). However, it is not necessary to introduce a special notation for the existential quantifier. For to say that there is something that has a particular property is simply to deny that everything fails to have the property. Thus our example could also be expressed, using only Frege's universal quantifier, as follows: $\sim(x) \sim$($x$ is a number that is $<2$).

This completes our sketch of Frege's logical notation and how it represents conceptual content. Frege needs not only a logical notation, but also a technique for evaluating inferences. His logic includes, in addition to a notation, a list of primitive logical laws and a rule that licenses the inference of one statement from others, that is, a rule of inference. In Part I of *Begriffsschrift,* Frege provides a list of primitive logical laws and a rule of inference from which, he believes, all truths of logic can be derived. The introduction of his logic is followed by an extensive demonstration of what the logic can do. All inferences that can be shown to be valid inferences using the techniques of Aristotelian or propositional logic can also be shown to be valid using Frege's new logic. That is, when the premisses and conclusion are expressed in Begriffsschrift, it is possible to derive the conclusion from the premisses using only Frege's laws and rule of inference. In Part II, Frege shows how to derive many of these inferences in his logic.

But the results of Frege's logic are not limited to those of Aristotelian and propositional logic. Part III of *Begriffsschrift* is devoted to derivations of arguments belonging to the general theory of sequences. As we saw near the end of Chapter 2, there are arguments about sequences that seem to create a problem for Aristotelian logic. They seem to depend only on analysis of concepts—in particular they seem to require no appeals to the evidence of the senses or to truths about spatial relations for their justification. Thus they ought to be logically valid. Yet they cannot be identified as valid by Aristotelian logic. One example is:

> Anna is a direct ancestor of Charles
> Anna has brown eyes
> All children of brown-eyed people have brown eyes
> Therefore, Charles has brown eyes

The third premiss tells us that the property of having brown eyes is hereditary in this ancestral sequence. Frege's notation allows him to express the conceptual content involved in the general claim that a property, say $P$, is hereditary in an ancestral sequence, say of a relation $f$. To say that $P$ is hereditary in the $f$-sequence is to say:

$$(x)[P(x) \rightarrow (y)\,(f(x, y) \rightarrow P(y))]$$

Or, in stilted English:

> For any object that has property $P$, all immediate descendants of that object have $P$.

If we regard '$f(x, y)$' as expressing the claim that $x$ is a parent of $y$, and '$P(x)$' as expressing the claim that $x$ has brown eyes, then the displayed expression expresses our third premiss. Among the propositions Frege proves in Part III is one, proposition 81, that allows us to infer the conclusion of the above argument from its premisses.

This is an important result. For, as we also saw earlier, arguments that involve ancestral sequences, as this one does, have a special relation to certain arguments that seemed, before Frege's work, to be peculiar to the domain of arithmetic: arguments employing the principle of mathematical induction. Mathematical induction tells us that, provided a property can be shown to hold of 1 and provided it can be shown to be hereditary in the natural number sequence, then the property holds of all natural numbers. That is, if a property holds of the number 1 and is hereditary in the immediate-successor sequence, then it holds of all members of the sequence. Thus one of the tasks Frege originally set himself was to show that logic alone suffices to show the validity of certain inferences about properties that are hereditary in ancestral sequences. This is the accomplishment of Part III of *Begriffsschrift*.

Frege's aim is to show that all truths of arithmetic are derivable from logic alone. In Part III of *Begriffsschrift*, he has shown that a principle that is used in many proofs of arithmetic—the principle of mathematical induction—can be replaced by a principle about ancestral sequences that depends only on logical laws. This constitutes an important step forward in Frege's project. But it is still just a first step. The above argument that Charles has brown eyes depends on facts about Charles, Anna, and a particular ancestral sequence, Anna and her descendants. If we want to establish that Charles has brown eyes, we will need to establish all the premisses of the argument: that Anna has brown eyes, that Charles follows Anna in the ancestral sequence, and that having brown eyes is hereditary in the ancestral sequence in question. Similarly, a proof in which the principle of mathematical induction is used will depend not only on that principle, but also on claims that 1 has the property in question and that the property in question is hereditary in the natural number sequence. Frege must also be able to show how these sorts of claims—that 1 has a particular property, that a property is hereditary in the natural number sequence—can be derived using nothing more than definitions and logical laws. For this, he will need definitions of the number 1 and of the concept of number.

# 4 *Defining the numbers*

In *Foundations of Arithmetic,* Frege turns to the project of defining the number one and the concept of number. *Foundations* is not, however, the sort of book one might expect. One might expect him to build on the achievements of *Begriffsschrift*: to formulate his definitions in the new logical language and to follow the definitions with proofs of at least some of the basic truths of arithmetic from these definitions and logical laws. Yet Frege makes very little use of the *Begriffsschrift* symbols and machinery in *Foundations*. He does not express his definitions in his logical language, nor does he attempt to provide gapless proofs. Instead, *Foundations* contains, in addition to some general discussions of the project, an elaborate and extensive informal discussion of his strategies for defining the numbers and constructing his proofs. Why is this?

It is easy to see why Frege should devote some time, at this point, to a general discussion of the project. Although he makes some allusions to his project in *Begriffsschrift,* he does very little to motivate the project or to explain how he intends to achieve his aims. *Foundations* contains his first extended explanation (the explanation we examined in Chapter 2) of how the definitions and proofs he proposes to give can provide an answer to his question about the source of our knowledge of arithmetic. It is less easy to see why he felt the need to provide elaborate informal discussions of his strategies for defining the numbers. After all, Frege did not engage in such discussions of his *Begriffsschrift* definitions. These definitions were explained in a few sentences and followed, almost immediately, with a series of proofs in the logical language. Frege's definitions of the numbers, however, play a role in his project that is different from the role of the *Begriffsschrift* definitions. As a consequence the definitions of the numbers must satisfy different demands. To see why, it will help to consider one of the *Begriffsschrift* definitions.

In *Begriffsschrift*, Frege introduces the expression '*F* is hereditary in the *f*-sequence' as a short way to say that whenever one member of the *f*-sequence has *F*, anything that immediately follows it in the *f*-sequence has *F*. He expects his readers to see the connection between this explicitly defined relation and everyday talk about properties that are literally inherited (by offspring from parents). But his use of 'hereditary' is not meant to capture the content of our everyday use of the word. It is far from evident that, for a property to be hereditary in the everyday sense, it must be inherited by *all* offspring. Further, as Frege uses the term, but not as we use it in everyday life, properties of numbers can be said to be hereditary. Frege uses the word 'hereditary' not because his aim is to capture the content we associate with the everyday word but, rather, because a consideration of this everyday content will help us understand the technical content that he wants to introduce in his definition. Indeed, the expression Frege actually defines, '*F* is hereditary in the *f*-sequence', is not an everyday expression at all. The definition itself is purely stipulative.

The purpose to be served by the definitions of the numbers is different. If Frege's proofs are to show us that the truths of *our* arithmetic are analytic, his definitions of the numbers must, in some sense, capture the content associated with our everyday use of the number words. It will not do, for instance, to say that the number one is Julius Caesar. It seems evident that no study of this historical figure will tell us anything about the truths of arithmetic. But what content must be captured? And how are we to recognize whether this content is captured? These questions are not easily answered. Much of *Foundations* is devoted to a discussion of the requirements that must be satisfied by acceptable definitions and of the difficulties involved in satisfying these requirements. It is only after these requirements and difficulties are understood that we can appreciate why it is that Frege's definitions are correct.

In the preface to *Foundations* Frege introduces, with great fanfare but little explanation, three fundamental principles that are to guide his enquiry. They are,

> always to separate sharply the psychological from the logical, the subjective from the objective;
> never to ask for the meaning of a word in isolation, but only in the context of a proposition;
> never to lose sight of the distinction between concept and object. (*FA*, p. x)

As we shall see, these principles play an important role, not only in Frege's procedures, but also in his conception of the project. Their significance, however, will become apparent only after we have seen how they function in his arguments.

In addition to these fundamental principles, Frege offers some specific requirements that definitions of the numbers must satisfy if they are to capture the requisite content. One of these is that the numbers, as defined, 'should be adapted for use in every application made of number' (*FA* §19). That is, given an acceptable definition of the number one, it must be true (failing new astronomical events) that the Earth has one moon. Moreover, although conceptual content is not explicitly mentioned in *Foundations,* Frege also makes clear that our everyday inferences must survive. For everyday purposes we are entitled to infer, from the claim that the Earth has one moon, that there is something that is a moon of the Earth. The introduction of a definition of the number one must not prevent us from making this inference. The definition should also enable us to fill in some of the gaps of the everyday inference—something that cannot be accomplished if our definition tells us that the number one is Julius Caesar. The truths and inferences that must be preserved by Frege's definition include not only our everyday applied inferences and truths, but also those of pure arithmetic. For the aim is 'to arrive at a concept of number usable for the purposes of science' (*FA* §57). Given acceptable definitions, it must be false that 0=1; true that 2 is the successor of 1, etc. These requirements, Frege thinks, must be satisfied by any acceptable definitions of the numbers.

But is it sufficient to satisfy these requirements? It may seem that it is not. If the number one and the concept of number are, as many writers think, simple non-logical notions, then the content we associate with a number word is simple. Consequently, any complex definition of the numbers from logical terms, that is any definition of the sort that Frege wants to give, will be incorrect. Thus one of his aims is to show that these apparently simple, well understood, notions are not well understood at all. His strategy is to argue that the views expressed in the discussion of these notions—not only in elementary textbooks, but also in the writings of mathematicians and philosophers—are either confused or incorrect. For these views do not satisfy the requirement that our everyday truths and inferences be preserved. Frege's examination of the defects of these views reveals, more clearly, the content that must be captured by acceptable definitions.

The first view that Frege considers, a view that is held by many of the writers he discusses, is that numbers are objective properties of external objects. It is not difficult to see the appeal of this view. Many of the everyday judgements in which number words are involved are judgements about the external world—for example, that the Earth has one moon. How is the content of this judgement to be understood? If one assumes the correctness of Aristotelian logic, as most of the writers Frege discusses did, then it should be understood as having subject/predicate form. If we look at the words used to express the judgement, it seems reasonable to assume that 'the Earth' is the subject and 'one' belongs to the predicate. Moreover, there is no obvious Aristotelian analysis on which 'one' is the subject of the statement. And, finally, what is predicated of the Earth seems to be something objective; something that is true or false of the Earth independent of any particular person's ideas or psychological states. It thus does not seem unreasonable to suspect that all the number words (including 'one') signify some sort of objective property that can hold of everyday physical, or external, objects. On this view, number is a property of objects in the way that colour or weight is a property of objects.

But such a view, Frege argues, cannot be correct. He notes that a person who is handed an object and asked to find its weight will know what to investigate. Moreover, the question will have a determinate answer. Were he to ask the same question of two people, he would expect to get the same answer. A comparable question about number, Frege argues, is very different. Suppose he handed someone a pile of cards and asked her to find its number. In this case we would not expect her to know what to investigate. Rather, we would expect her to ask for further information. For the question could be understood as:

What number of cards have I handed to you?
What number of complete packs of cards have I handed to you?

Or even,

What number of piles have I handed to you?

Each of these questions has a determinate answer and the answers may be different. It could be, for example, that the correct answer to the first question is '104'; the correct answer to the second is '2'; and the correct answer to the third is '1'. Given that the question 'What is the number of this?' can be understood in three different ways, each of which has a different answer, there cannot be a determinate correct answer to this question. A particular number, Frege says, cannot be said to belong to the pile of cards in its own right. Rather, it seems to belong to the pile 'in view of the way in which we have chosen to regard it'. (*FA* §22) The correct answer to Frege's question seems to depend on whether we regard the pile as a pile of cards or a pile of packs of cards. In contrast, the correct answer to the question about the weight of the pile remains the same no matter whether we regard it as a pile of cards or as a pile of packs of cards.

After raising this objection to the view that number is a

property of external objects just as weight and colour are, Frege goes on to raise related objections by focusing on a particular number, the number one. If the number one is a property of external things, what can account for the fact that one pair of boots is the same as two boots? He considers a variety of proposals for what oneness might be. Among these are that predicating oneness of something involves identifying it as an isolated, self-contained, or undivided object. But no such account will fit with our everyday judgements. For if correct predication of oneness requires the identification of something as isolated or undivided, then it would be wrong for me to identify a pair of boots discarded hastily so that each boot is left in a separate room as *one* pair of boots.

The moral of all these examples is that numbers are not objective properties of external objects. How are numbers related to objects? Frege's arguments may seem to show that a number cannot be a property of an object in its own right but, rather, can only be a property of an object under a description. In order to answer the question about what number should be ascribed to a pile of cards, we need to know whether to count cards, packs, or piles. In order to determine what number should be ascribed to some boots, we need to know whether to count boots or pairs of boots. One might suppose, then, that numbers are subjective properties; that a number belongs to an object in virtue of how we choose to regard it. Indeed, it may begin to seem as if, depending on our way of regarding the things in question, it can be correct to ascribe just about *any* number to *any* things. A boot, after all, could be regarded as consisting of two boot-halves, three boot-thirds, etc.

But the view that numbers are subjective properties also, Frege argues, fails to fit our everyday judgements. If number is a subjective property, then one answer is as good as another—the number a particular object has for me need not be the same number the object has for you. When a person is asked for the number of something in an everyday context, however, she cannot simply choose any number at all. Indeed, the situation Frege describes—a situation in which he hands someone a pile of cards and asks 'what is the number of this?'—will not arise in an everyday context precisely because it has no determinate answer. Anyone who receives a serious request of this sort will also be supplied with the requisite description. And if this description is supplied, she is not free to choose any number at all: there is a correct answer. Someone who is handed a pile of cards and asked how many cards are in the pile will be expected to count the cards. If there are 52 cards in the pile and she replies 'one' because she is regarding the pile as consisting of complete packs, she will have given an incorrect answer. Because there is a correct answer, number is not subjective.

Nor is the question about number a psychological one. As we have seen, one of Frege's fundamental principles is, 'always to separate sharply the psychological from the logical, the subjective from the objective' (*FA*, p. x). Psychology is certainly relevant to a concern with what might have led us to believe a particular answer is correct. Psychology is not, however, relevant to Frege's concern with correct answers and with what justifies them. The correct answer is not dependent on how someone regards the pile. It is irrelevant whether she regards the pile as consisting of packs or of cards—in either case, the correct answer is that there are 52 cards. Thus we do not need any knowledge of psychology (or of anyone's psychological state) to determine what counts as a correct, or incorrect, answer to the question. Numbers, Frege concludes, cannot be subjective properties of external objects.

At this point it may begin to seem that it is futile to attempt to satisfy the constraints that, on Frege's view, must be satisfied by acceptable definitions. Number words are used in assertions about the everyday physical world. Yet they clearly cannot be everyday physical objects like the Earth nor, as we now see, can they either be objective or subjective properties of these objects. But what other options are available? Instead of giving a direct answer to this question—instead of identifying a particular sort of entity that number words name—Frege analyses the role these words play in certain everyday statements about the world. This approach to the question may seem rather indirect, but it is exactly the sort of approach suggested by the second of his fundamental principles, 'never to ask for the meaning of a word in isolation, but only in the context of a proposition' (*FA*, p. x). Indeed, this principle, which contemporary philosophers call the 'context principle', is at work in the above arguments that numbers cannot be properties of physical objects. For one of his central arguments that numbers are not properties of physical objects came from a consideration of what sort of statements about numbers are correct. Frege argues that, if I were simply to indicate a physical object and say 'the number of this is 52', my statement would be neither correct nor incorrect. Only upon being supplied with a concept (for example: *cards* or *packs of cards*) are we entitled to say whether or not the statement is correct.

When Frege turns, next, to the issue of how number is to be defined, he does not consider numbers in the abstract but, rather, the content of statements about numbers. The statements on which he focuses, which he calls 'ascriptions of number', are statements of the following sort:

The pile contains 52 cards
The Earth has one moon
Venus has 0 moons
The King's carriage is drawn by four horses

These statements, he argues, are assertions about concepts.

As Frege realizes, it is tempting to think of concepts as something psychological. But his notion of concept is a logical, not psychological notion. And its importance is indicated by the third of Frege's fundamental principles, 'never to lose sight of the distinction between concept and object' (*FA*, p. x). The concept/object distinction comes from his new logic, on which the simplest statements can be broken down into function and argument. We can replace 'the Earth' in the sentence

The Earth is a planet

with a variety of different names, resulting in such sentences as:

Venus is a planet.

and

Mars is a planet.

What is invariant in all the above statements is a function that takes one argument. This sort of 1-place function—something that can be asserted of an object—is a concept. This notion of concept does not belong to psychology at all. There is nothing psychological about the claim that Venus is a planet. To say that Venus is a planet is not to say anything about any person's ideas about Venus but to say something objective about Venus. Concepts, and hence assertions about concepts, are not subjective or psychological.

Frege illustrates this with the following example,

All whales are mammals.

This is clearly an objective and non-psychological statement. What is it about? Although it may seem to be a statement about animals, no name of an animal appears in this statement. Indeed, even supposing a whale were before me, I would not be licensed to infer, from this statement alone, that it is a mammal. In order to make this inference, I would need an additional premiss: the premiss that it is a whale. On Frege's view, the statement in question is a statement about concepts—the concept *whale* and the concept *mammal*. One way to see this is to consider its logical regimentation. To say that all whales are mammals is to say that, no matter what object we choose, say *x*, if *x* is a whale, *x* is a mammal. That is, the logical regimentation of this statement will be:

$(x)\,(Wx \rightarrow Mx)$

We can see what this statement is about by looking at what symbols in the regimentation function as names. The symbols in the above regimentation include parentheses, a quantifier '$(x)$', a variable '$x$' and a logical connective '$\rightarrow$' as well as two upper-case letters '$W$' and '$M$'. Only the two upper-case letters stand for particular, non-logical things: the concept *whale* and the concept *mammal.* We have, then, an entirely objective non-psychological statement about concepts.

We are now in a position to understand Frege's analysis of ascriptions of number. Let us begin with a simple example. To say that Venus has 0 moons is to say that, for anything we choose, no matter what it is, it is *not* a moon of Venus. That is, if we use the symbol '*V*' to represent the concept of being a moon of Venus,

$$(x) \sim Vx.$$

As with our analysis of 'all whales are mammals', this analysis indicates that to say that Venus has 0 moons is to make a claim about a concept: *moon of Venus*. Similarly, to say that the Earth has exactly one moon is to say something about the concept, *moon of the Earth*. It is to say that there is something, say *x*, which is a moon of the Earth and that, for anything we choose, if it is a moon of the Earth, then it is identical to *x*. Suppose we introduce a simple symbol, say '*E*', for the expression 'is a moon of the Earth'. Using this abbreviation, the statement can be represented in logical notation as follows,

$$(\exists x)\,(Ex\ \&\ (y)\,(Ey \rightarrow y=x)).$$

Of course, ascriptions of number need not be true. We might, for example, say that the Earth has exactly two moons. This statement, also, can be understood as a statement about the concept, *moon of the Earth.* To say this is to say that there are distinct things, say *x* and *y*, both of which are moons of the Earth, and that everything which is a moon of the Earth is either identical to *x* or identical to *y*. This can be represented as follows,

$$(\exists x)\,(\exists y)\,(Ex\ \&\ Ey\ \&\ {\sim}x = y\ \&\ (z)\,(Ez \rightarrow z=x \vee z = y)).$$

This strategy can be used for any ascription of number (including, for example, that there are 52 cards in a pack), although the sentences in question become very complicated as the numbers get larger. The moral is that ascriptions of number are statements about concepts.

This construal of the content of ascriptions of number lends plausibility to Frege's claim that one and the concept number can be defined from purely logical terms. For we have just seen that the contribution that the word 'one' makes to the statement that there is exactly one moon of the Earth can be expressed in purely logical terms. Moreover, it may seem that we now have a way to define the number words. Number words, it may seem, are simply abbreviations of complicated quantifier expressions. This is misleading, however. One hint of the mistake is that the complicated quantifier expressions do not actually replace number words. In the above translation of 'the Earth has exactly two moons', the complicated quantifier expression does not replace 'two'. Rather, it replaces the more complex expression, 'has exactly two'. In order to see why we do not yet have definitions of the numbers, we need to consider, again, the nature of Frege's project. The above discussion shows us how to replace all number words that appear in ascriptions of number with expressions of logic. But Frege wants to provide a foundation for arithmetic, not just for its applications. In addition to preserving the everyday applications of number, Frege's definitions are meant to make it possible to prove the truths of pure arithmetic from logical laws. For example, for definitions of '0' and '1' to be acceptable, it must be provable that 0 is not identical to 1. The claim that 0 is not identical to 1, however, does not seem to be an ascription of number or an assertion about a concept. How are the statements of pure arithmetic to be understood?

As the example about 0 and 1 indicates, many of the statements of pure arithmetic are identities. Identity statements are about objects. If we can say that 1 is identical (or not) to 0, then 1 must be an object. More generally, all numbers are objects. It may seem that this conclusion will not fit with the earlier account of ascriptions of number. Frege writes,

Now our concern here is to arrive at a concept of number usable for the purposes of science; we should not, therefore, be deterred by the fact that in the language of everyday life number appears also in attributive constructions. That can always be got round. For example, the proposition 'Jupiter has four moons' can be converted into 'the number of Jupiter's moons is four'. (*FA* §57)

Frege's task, then, is to provide definitions of the numbers on which they are self-subsistent objects and from which it is also possible to derive the requisite applications. In particular it must be derivable—from the definition of the number one along with the claim that the number of moons of the Earth is one—that there is something which is a moon of the Earth and that every moon of the Earth is identical to that thing.

At this point, however, we still do not have the requisite definitions. What is required? Since it is one of Frege's fundamental principles never to ask for the meaning of a word in isolation, but only in the context of a proposition, he begins by identifying the propositions that are of import in constructing the definitions. As we have seen, Frege has already argued that numbers must be objects. Thus the propositions of import include identity statements. If '*a*' and '*b*' are names of objects then, Frege claims, 'we must have a criterion for deciding in all cases whether *b* is the same as *a*' (*FA* §62). This is not to say that we can always apply the criterion. We may not have the computing resources available to determine whether a complex equation is true or false. Nonetheless, the definitions must provide criteria that determine whether it is true or false. Supposing, then, that *a* and *b* are numbers, when are they the same?

One strategy for determining whether, for example, there is the same number of plates on a table as forks, is to line up the plates and forks in pairs; to put one plate to the right of each fork (as they are lined up in a typical table setting). If this can be done and there are no plates or forks left over, then there is the same number of plates and forks. Indeed, a version of this procedure can be used to determine whether there is the same number of plates as forks even if it is not actually possible to line the pairs up physically. So long as there is a way of associating plates and forks so that each plate is paired with a unique fork and there are no leftover plates or forks, we say that there is a *one–one correspondence* between plates and forks. In general, for any concepts *F* and *G*, if there is a one–one correspondence between *F*s and *G*s, then there is the same number of *F*s as *G*s. Frege begins his introduction of numerical identity by suggesting that the content of the proposition

> The number which belongs to the concept *F* is the same as that which belongs to the concept *G*

is that there is a one–one correspondence between *F*s and *G*s. If there is a one–one correspondence between *F*s and *G*s, *F* and *G* are said to be 'equinumerous'.

This is not enough, however, to give us the content of all identity statements involving numbers. For according to Frege's requirement, if the number one is an object, then there must be a criterion for deciding, for any object—including, for example, Julius Caesar—whether or not it (he) is identical to the number one. Now it seems obvious that Julius Caesar is not a number at all, let alone the number one. This illustrates a problem with using the above analysis to define the number one. Frege requires a definition of the number one to provide a criterion that decides, among other things, whether Julius Caesar is the number one. But the definition of numerical identity in terms of one–one correspondence provides such a criterion only if the expressions on each side of the identity sign are of the form 'the number that belongs to the concept *F*'. If Julius Caesar is not a number, then he is not the number one. It is, then, one of the burdens of a definition of the number one to provide criteria that rule out taking Julius Caesar to be the number one. This burden cannot be met, however, by the sort of definition of numerical identity described above.

Because it works only for a specific sort of numerical identity statement, the above explanation will not provide the requisite identity criterion. And, since it is part of Frege's project to define the concept of number, it will not do simply to *assume* that one is a number and then identify properties that distinguish one from the other numbers. Frege might choose to begin with a definition of the concept of number. But he does not. Instead he attempts to identify each number as an object of the appropriate sort. His strategy for defining each number is to identify an object that is associated in the appropriate way with (belongs to) each concept that holds of exactly that many objects. In order to define zero, for example, Frege needs to identify an object that belongs to each concept that holds of no objects, that is, an object that belongs to such concepts as *moon of Venus*, etc. In addition, it must be evident that this object is not identical to Julius Caesar. It must be possible, as well, to derive the appropriate applications from the definition. For example, from the definition of 0, and the claim that the number 0 belongs to the concept *moon of Venus*, it must be possible to derive the statement that Venus has no moons.

What objects are associated with concepts? One obvious answer is that a concept is associated with the objects that fall under it (or, the objects of which the concept holds). Since each of us has a heart—that is, since the concept *has a heart* holds of each of us—we each are associated in this way with the concept *has a heart*. An object that is associated with a concept in this way is said to be *in the extension of the concept*. Since each animal that has a heart also has blood vessels and vice versa, an animal is in the extension of *has a heart* just in case it is in the extension of *has blood vessels*. Thus we can say that these concepts have the same extensions. Since sameness, or identity, is a relation that holds between objects, extensions must be objects.

But what is an extension of a concept? Frege has surprisingly little to say about this. Indeed, when he summarizes the achievements of *Foundations* in Part V, which is labelled 'Conclusion', he merely remarks

In this definition the sense of the expression 'extension of a concept' is assumed to be known. (*FA* §107)

Were Frege introducing the term 'extension' for the first time, this would seem to be a cheat. However, this was a traditional logical term, first used, not in *Foundations*, but in an influential seventeenth-century work, *La Logique ou l'Art de penser*. The term was widely used by logicians in Frege's time (and continues to be used today). Frege means to rely on his readers' familiarity with the traditional assumption that there must be some object (an extension) associated with all concepts that hold of exactly the same things. Moreover, he believes that this assumption is fundamental to logic; the notion of extension, he claims, must simply be taken as a primitive logical notion. Consequently, Frege feels he need offer no definition of 'extension'. In the context of these views, the definitions of numbers that Frege proposes to give—definitions on which numbers are extensions of concepts—will also solve the Julius Caesar problem. Since we know what sort of thing extensions are, we know that Julius Caesar is not an extension of a concept. Hence, given a definition on which the number one is the extension of a particular concept, Julius Caesar is not the number one.

As we will see, Frege's strategy is mistaken. The introduction of the notion of extension of a concept leads to disaster—an inconsistency in his logic. Although Frege goes on to say, in *Foundations*, that he attaches no decisive importance to bringing in the extensions of concepts, he later came to see that the introduction of extensions of concepts is of crucial importance for his ability to offer the appropriate sorts of definitions. We will conclude our examination of *Foundations* by seeing how he exploits the notion of extension to complete his definitions.

Frege bases his definitions on the insight that the same number belongs to concepts that are equinumerous. That is, if the *F*s can be put in one–one correspondence with the *G*s, then the same number belongs both to *F* and *G*. Suppose we want to identify the number that belongs to a particular concept, *G*. The aim, then, is to identify an object that is associated with *G* and with every concept that is equinumerous with *G*. Although Frege does not express it this way, we can also say that the aim is to identify an object that is associated with the **extension of *G*** and every **extension** that is equinumerous with *G*. To see why this is so, it is important to remember that the *F*s are simply those things that are in the extension of *F*. Hence to say that **the *F*s** can be correlated one–one with the *G*s is to say that **the**

**members of the extension of *F*** can be correlated one–one with the *G*s (or with the members of the extension of *G*). This is important because, in order to avoid addressing some technical difficulties with Frege's actual introduction of his definitions, it will help to talk about extensions rather than concepts.

Let us now consider a particular example, the concept *is a letter of the word 'Mars'*. The members of the extension of the concept *suit in a pack of cards* can be correlated one-one with the letters of the word 'Mars'. Thus:

> the extension of the concept *suit in a pack of cards* is equinumerous with the concept *letter of the word 'Mars'*.

Indeed, this works for any extension of a concept that holds of exactly four objects. Also:

> the extension of the concept *angles in a parallelogram* is equinumerous with the concept *letter of the word 'Mars'*.

The aim, if we are to find the number of the concept *letter of the word 'Mars'*, is to identify an object that is associated with each of these extensions—that is, with each extension that is equinumerous with the concept *letter of the word 'Mars'*. As this indicates, what distinguishes all these extensions is that they all fall under a particular concept:

> is equinumerous with the concept *letter of the word 'Mars'*.

Or, in other words, each of these extensions is in the extension of the above concept. The extension of this concept, then, that is:

> the extension of the concept is equinumerous with the concept letter of the word 'Mars'

is an object that satisfies the criteria with which we started.

Frege defines the number that belongs to *F* as the extension of *is equinumerous with the concept F*. One way to think about this definition, as we saw in the above example, is that it amounts to saying that a number, say 4, contains the extensions of all concepts under which exactly 4 objects fall. One of Frege's desiderata has now been satisfied: we have the means for formulating definitions on which numbers are objects. But this is not enough. Frege wants to show that all the truths of arithmetic are derivable from logical laws and definitions alone. His next step is to make it plausible that this is so.

For this purpose it is important to see that there can be definitions of the individual numbers that rely only on logical notions. For example, both

*is the author of* Foundations

and

*is a moon of the Earth*

are satisfied by exactly one object, and any concept that is equinumerous with either is equinumerous with both. Thus the concepts

*is equinumerous with the concept is the author of* Foundations

and

*is equinumerous with the concept is a moon of the Earth*

have the same extension: the extension whose members are all extensions of concepts that hold of exactly one object. Hence either of these expressions could be used to define the number one. But neither of these proposed definitions will be useful for Frege's purposes. The problem is that it is not a logical truth that there is a unique author of *Foundations*. Nor is it a logical truth that there is a unique moon of the Earth. Were Frege to use either of these definitions in his proofs about properties of the number one, the proofs would have empirical, as well as logical, premisses. Thus another part of Frege's strategy is to provide definitions of the individual numbers from purely logical notions. For this purpose it is helpful to begin not with the number one, but with the number zero.

To define zero, Frege needs to identify a concept that can be shown, using logical laws alone, not to hold of any object. Consider the property that holds of an object just in case it is identical to itself. Each object is identical to itself. Moreover, no information about an object—no additional fact—is required to see that the object is identical to itself. It is, consequently, a logical truth that each object is identical to itself or that the concept *is self-identical* holds of each object. It is thus also a logical truth that the concept *is not self-identical* does not hold of any object at all—nothing is in its extension. That is to say, the number zero belongs to the concept *is not self-identical.* Frege defines zero as the extension of *is equinumerous with the concept is not self-identical.*

To define one, Frege needs a concept that can be shown to hold of exactly one object. Since his definition of zero picks out a unique object, the concept *is identical to zero* holds of exactly one object. A concept that is equinumerous with *is identical to zero* will have exactly one object in its extension. He defines the number one as the extension of *is equinumerous with the concept is identical to zero.* That is, the number one is the extension that has in it all one-membered extensions. Once zero and one have been defined, we have two objects, which can be used to construct a concept whose extension has two members: *is identical to zero or is identical to one.* The extension of the concept of being equinumerous with that concept, the extension that contains all two-membered extensions is, by Frege's definition, two. This strategy can be continued indefinitely.

Frege can also define the relation that holds between a number and its successor in purely logical terms. A number *n* is the successor of *m* just in case:

> there exists a concept *F*, and an object falling under it *x*, such that the number which belongs to the concept *F* is *n* and the number which belongs to the concept 'falling under *F* but not identical with *x*' is *m*.

To see how this works, consider the claim that 4 is the successor of 3. To show that this is so we need to find a concept, and an object falling under it, that satisfy the above description. Let us consider, again, the concept *is a letter of the word 'Mars'*. One of the objects that falls under it is the letter '*a*'. We now have everything we need to show that 4 is the successor of 3. The number that belongs to the concept in question is 4. The number that belongs to the concept *is a letter of the word 'Mars' but is not identical with 'a'* is 3.

These definitions seem to be correct, but do they really capture the basic notions of *our* arithmetic? And how can we tell?

Frege acknowledges that it will not be evident at first that his definitions are correct because, he says, we 'think of the extensions of concepts as something quite different from numbers' (*FA* §69). His next move, one would expect, will be to argue that this is due to our not having thought the matter through—that on more careful consideration we will see that numbers really *are* extensions of concepts. But he does not make such an argument and, indeed, later in *Foundations* says that he attaches no decisive importance to bringing in the extensions of concepts at all (*FA* §107). This attitude seems very peculiar if we assume, as we certainly seem to do in our everyday use of the number words, that we are already using these words to talk about particular objects. For, if so, these particular objects either are or are not extensions of concepts. If they are extensions of concepts, it is *essential* for Frege to bring in extensions when he defines the numbers; if they are not it is *wrong*. On this natural assumption, Frege's procedure is simply wrong; and his definitions may well be incorrect.

What are we to infer? We might simply infer that, when Frege actually gets to the point of defining the numbers, he fails to notice the consequences of his procedure. But it is not very plausible that he is making this sort of mistake. Suppose that his aim is to come up with a precise description that picks out the object that we mean, all along, to be talking about when we use, for example, the term '1'. If so, it is simply inexplicable that he never actually tries to show that his definition satisfies this requirement. Thus it seems reasonable to infer that this is not his aim. But, if he assumes that, in our everyday use of the term '1', we *are* talking about a particular object, then it is inexplicable that he should want to define '1' without picking out this object. The only conclusion to draw, if we want to avoid regarding Frege's procedure as inexplicable, is that he does not believe that our numerals already pick out particular objects. Moreover, a number of Frege's other remarks suggest that this conclusion is correct. For example, he says explicitly that the concept number has not yet been fixed for us (*FA* §63). Yet, if our numerals already pick out particular objects, then it would be a determinate fact that each object either is or is not one of these—the concept of number would already have been fixed.

But to draw this conclusion seems simply to relocate the problem. If the numerals do not already designate particular objects then how can sentences in which they appear express truths? If '1' does not already designate a particular object, how can we regard it as true that $1+1=2$? If, as Frege says, the concept of number is not yet fixed, how can we regard it as true that every number has a successor? What licenses Frege to hold onto the view that our everyday statements of arithmetic express truths?

Although Frege does not address these questions in *Foundations*, some answers are suggested in later writings about his project. In order to see how these answers work, it is important to note, first, that there are circumstances in which we regard certain terms as not having fixed meaning yet, at the same time, regard some sentences containing them as expressing truths. One of the aims in developing a scientific field is to settle on appropriate concepts. In the early stages of research in some field, the formulations of hypotheses that are tested contain terms that do not yet have fixed meaning. If the hypotheses pass the tests, they are regarded as truths. Were arithmetic a science in its early stages, it would not seem absurd to suggest that the concept of number is not yet fixed and, at the same time, to regard statements about numbers as expressing truths. Of course, it is difficult to regard arithmetic as a science in its early stages. For arithmetic is the simplest, and most well understood, science available. Yet a consideration of Frege's characterization of his epistemological project suggests that this is precisely the attitude he wishes us to take towards arithmetic.

Frege's epistemological project carries with it a demand that science be held to new standards of rigour. The identification of the foundations of a science, on his view, requires all concepts of the science to be defined from primitive terms and all arguments to be replaced by gapless proofs from primitive truths. To introduce such definitions and proofs is, to use Frege's terminology, to systematize the science—the definitions and proofs constitute a system. And the pre-systematic (in Frege's sense) stages of a science should be regarded as early stages. Science comes to fruition only in a system. Frege means not only to introduce a new general requirement—that each science must, ultimately, be systematized—he also means to introduce the first systematic science, a systematic science of arithmetic. The pre-systematic concept of number, on Frege's view, is not yet fixed; nor, consequently, does our pre-systematic use of the numerals pick out particular objects. Yet this is not to say that pre-systematic arithmetic, or other pre-systematic sciences, are of no interest—indeed, we cannot begin with systematic science. Without pre-systematic science there would be no science to systematize. Moreover, there could be no advances in pre-systematic science if the practitioners did not regard their statements as having truth-values.

Frege's attitude towards pre-systematic science is evident in his discussion of Karl Weierstrass (1815–97), one of the greatest mathematicians of the nineteenth century. Weierstrass, Frege says, lacked the ideal of a system and, as a consequence, had only a very hazy notion of what number is. How, then, can we account for Weierstrass's prodigious accomplishments? These accomplishments, of course, involved drawing inferences from statements that he regarded as truths. And, as we saw in Chapter 2, it seems possible to draw inferences from statements that one does not entirely understand. Even if we do not know what nudibranchs are, if we know they are mammals and that all mammals are vertebrates, we can infer that nudibranchs are vertebrates. What is basic to *our* arithmetic is not a group of objects to which our numbers refer but, rather, a group of statements that we regard as the basic truths of our arithmetic, both pure and applied. Thus Frege need not convince his readers that his definition of '1' picks out something they already recognize independently of these statements. He needs, rather, to convince his readers that his definitions preserve what they regard as the basic truths of arithmetic. It must be true, given Frege's definitions, that 0 is not identical to 1; that each number has a unique successor; that, if the number one belongs to a concept, there exists an object which falls under the concept. It is one of Frege's aims to introduce a systematic science of arithmetic in which it is possible to reproduce, among other things, Weierstrass's proofs.

After introducing his definitions, Frege offers brief sketches of how these, and several other basic truths of arithmetic, can be proved using his definitions and logical laws. These are only sketches, of course. The language for expressing gapless proofs is Begriffsschrift. But these sketches, both of the definitions and the sort of proofs that can be constructed using them, accomplish the stated aim of *Foundations*: Frege has made his case that it is probable that the truths of arithmetic can be derived from logical laws and definitions alone. The next step, it seems, is to give the official Begriffsschrift proofs of the truths of arithmetic.

Before Frege is able to take this step, however, he needs to introduce a new version of Begriffsschrift. One reason is that his strategy for defining the numbers requires a logical notation for extensions of concepts. And while Frege claims, in *Foundations*, that the notion of extension of concept is a primitive logical notion, there is no notation for, or even mention of, extensions of concepts in *Begriffsschrift*. It is not only the notation that needs to be changed. The discussions of *Foundations* highlight a number of difficulties with his *Begriffsschrift* explanations of some fundamental logical notions. One important piece is simply missing. The notion of concept, which plays a central role in the arguments of

*Foundations*, is rarely mentioned in *Begriffsschrift*. Other characterizations of fundamental logical notions are different from, and incompatible with, the *Begriffsschrift* characterizations. One of Frege's next tasks is to provide a more comprehensive explanation of the notions underlying his logic.

# 5 *The reconception of the logic: I*

## 'Function and Concept'

As we have seen, one of Frege's fundamental principles is 'never to lose sight of the distinction between concept and object' (*FA* p. x). One of his central insights in *Foundations* is that ascriptions of number are statements about concepts. But what is a concept? It is remarkable how little Frege says about this in *Foundations* and the earlier works. Given the function/argument analysis presented in *Begriffsschrift*, it seems reasonable to suppose that a concept is some sort of function. But Frege says almost nothing about concepts in *Begriffsschrift*. In particular, there is no explanation of the connection between the notion of concept and the general, logical notion of function in *Begriffsschrift*. The reason, one might suspect, is that Frege realized the significance that the notion of concept has for his project only afterwards. But, in *Foundations*, where this notion plays a prominent role, the issue of whether concepts are functions is not raised explicitly. There is, of course, strong evidence that the role played by concepts in the statements Frege analyses in *Foundations* is the same as the role played by functions in the analyses of *Begriffsschrift*. But he does not make much of an attempt to distinguish his notion of concept from the Aristotelian notion of concept. It is not evident, for example, that anything in the Aristotelian view conflicts with Frege's statement that 'With a concept the question is always whether anything, and if so what, falls under it' (*FA* §51). In spite of the fact that the purpose of *Foundations* is to make it plausible that the truths of arithmetic are really logical truths, the general logical notion of function is virtually ignored there. The discussions of functions in *Foundations* concern, almost invariably, either the familiar mathematical notion of function or particular mathematical functions.

An examination of his early remarks about functions and concepts suggests that this is not an inadvertent, but easily rectified, omission. For these remarks display a nest of difficulties. One of these is that his only characterization of functions, the characterization in *Begriffsschrift,* seems to provide no support for the view in *Foundations* that there is a fundamental distinction between concept and object. In fact, according to the official view of *Begriffsschrift*, there is no fundamental distinction between function and argument.

As we saw earlier, Frege introduces the notion of function, in *Begriffsschrift*, by asking the reader to consider replacing the word 'hydrogen' in

> Hydrogen is lighter than carbon dioxide

with various other words, including 'oxygen' and 'nitrogen'. These replacements give us different sentences,

> Oxygen is lighter than carbon dioxide.

and

> Nitrogen is lighter than carbon dioxide.

This amounts, he claims, to considering *is lighter than carbon dioxide* as a function and considering hydrogen, oxygen, and nitrogen as its arguments. The particular function in question is a concept—that is, it is a function that takes one argument and it is something that can be asserted of an object. Thus far, the description accords with the remarks about concepts from *Foundations*. Concepts are distinguished from objects by the fact that one of the important questions we ask about a concept is whether anything, and if so what, falls under it. Each of the above sentences appears to be part of an answer to this question where the concept in question is *is lighter than carbon dioxide*. In contrast, it makes no sense to ask this question about an object. It seems reasonable to infer from this that what distinguishes concepts from objects is that a concept is a sort of function and that 'argument' is Frege's *Begriffsschrift* term for 'object'. However, while this is suggested by the above examples, on the official *Begriffsschrift* view it is impossible to draw a principled distinction between functions and non-functions.

The problem is that such words as 'hydrogen' and 'carbon dioxide' are not the only sort of expressions that can be viewed as replaceable in sentences in which they appear. After giving his characterization of function Frege notes in an aside that we can view not only the predicate, but also the subject of our sentence, as designating a function. Instead of leaving the predicate unchanged and replacing the subject of the sentence, we can leave the subject unchanged and replace the predicate by other predicates. The sentences,

hydrogen is lighter than oxygen

and

hydrogen is heavier than carbon dioxide

both come from our original sentence by replacing the original predicate. We can consider not only *is lighter than carbon dioxide* but also *hydrogen* as a function. Consequently, we can analyse the content of the sentence 'hydrogen is lighter than carbon dioxide', not only as it is analysed above, but also as the value of the function *hydrogen* for the argument *is lighter than carbon dioxide*. Any significant constituent of a sentence can, it seems, be regarded either as function or as argument. The function/argument distinction, he says in *Begriffsschrift*, has nothing to do with conceptual content.

This characterization of functions cannot be reconciled with the admonition never to lose sight of the distinction between concept and object. The view of *Foundations* seems to be that a concept is a particular sort of function, and that the sort of argument it takes is an object. Moreover, when Frege claims that ascriptions of number are statements about concepts and that numbers are objects, he means to be telling us something about the correct logical analyses of ascriptions of number and statements about numbers. But, according to the *Begriffsschrift* view, such claims cannot have this significance. For we are free to regard *any* expression as a function-expression. If so, there is no basis for saying that numbers are objects, rather than concepts. There can be no principled distinction between concept and object. This is a serious problem, given that the existence of a principled distinction between concept and object seems essential to the conception of number that Frege introduces and defends in *Foundations*.

The arguments of *Foundations* made it evident that Frege needed a new characterization of function. Moreover, the *Foundations* characterization of the concept/object distinction proved to be too minimal to convince his readers. One of them, Benno Kerry, argued in a paper published three years after *Foundations* appeared, that the concept/object distinction was not absolute—that Frege's own characterization of the distinction allowed for objects that are also concepts. In the years following the publication of *Foundations*, he set himself the task of sorting out the basic notions underlying his new conception of logic. He began with the central notions of function and concept, for which he provided new characterizations in a shorter work entitled 'Function and Concept'.

'Function and Concept' begins with a look at familiar mathematical functions. Frege draws our attention to three expressions, '$2.1^3+1$', '$2.4^3+1$', and '$2.5^3+5$'. We can see that something is common to all three expressions. The first is an expression for the value that a particular function has for 1 as argument; the second is an expression for the value of the same function with 4 as argument; and the third is an expression for the value of the same function with 5 as argument. Frege writes (FC 6),

From this we may discern that it is the common element of these expressions that contains the essential peculiarity of function; i.e. what is present in

'$2.x^3 + x$'

over and above the letter '$x$'. We could write this somewhat as follows:

'2( )$^3$+( )'.

Frege makes use of these observations to introduce a new characterization of function-expressions: function-expressions are incomplete. A closer look at these observations can show us something about how Frege means us to understand this new characterization.

Since each of the expressions, '$2.1^3+1$', '$2.4^3+1$', and '$2.5^3+5$', is a name for the value a particular function has on some argument, it seems reasonable to suppose that an expression for this function is a constituent of each of these expressions. Thus, since none of them contains the letter '*x*', this letter cannot be part of the function-expression. But it will not quite do to regard the expression with empty spaces as the requisite function-expression. For this would not allow us to distinguish the function of one argument that Frege wishes to talk about in the above quotation from the two-argument function for which we would normally use the expression '$2.x^3+y$'. There is, it seems, no way simply to write out the function-expression on its own. Only in combination with argument expressions (either variables or number-names) can function-expressions actually be written out. A function-expression is incomplete. And the incompleteness of function-expressions exhibits a feature of functions as well. A function, on Frege's new characterization, is incomplete, in need of supplementation, or *unsaturated*.

Frege introduces this new characterization of functions, as he introduced his *Begriffsschrift* characterization, by drawing our attention to expressions and to their parts. But, in 'Function and Concept', unlike *Begriffsschrift*, he is careful to link his claims about expressions with claims about their content. Functions are to be distinguished from function-expressions. Function-expressions, like number expressions, have content. Different expressions, for example '7+5' and '6.2', can be expressions for the same number. There is nothing inherent in the *expressions* that explains this—the explanation must lie in the content, not only of the numerals, but also of the constituent function-expressions. Function-expressions are used to talk about functions, just as numerals are used to talk about numbers. The unsaturatedness of function-expressions reflects an unsaturatedness that is, Frege says, the essence of functions. And, he says, 'functions differ fundamentally from numbers' (FC 6).

The functions that Frege uses when he introduces the notion of unsaturatedness, as the above examples indicate, are all simple functions of arithmetic. But Frege's aim, as we know from his earlier work, is to use the familiar mathematical notion of function to introduce a more general, logical notion of function. It is not only mathematical functions that exhibit unsaturatedness, the essence of functions. Concepts, as well as mathematical functions, are unsaturated. Consider the content of '1 is less than 2'. We can understand this as predicating a concept, *is less than 2*, of 1. As with a function-expression, a concept-expression is somehow incomplete. The expression 'is less than 2' cannot be a real concept-expression because it is missing an indication of what supplementation is required. Nor can we obtain a concept-expression by supplying such an indication. Suppose we require the concept-expression to include a variable that indicates what supplementation is required—as in, for example, '*x* is less than 2'. Then the sentence '1 is less than 2' does not contain the requisite concept-expression, for part of the concept-expression (i.e. '*x*') does not appear in the sentence. The incompleteness of concept-expressions, like the incompleteness of function-expressions, reflects an incompleteness in their content. Concepts are unsaturated. And an understanding of unsaturatedness is what we need in order to grasp Frege's notion of objecthood. Frege writes, 'an object is anything that is not a function, so that an expression for it does not contain any empty place' (FC 18).

We now have a new, more general notion of function. This includes not only familiar mathematical functions but also concepts; not only concepts that we think of as applying to numbers, but *all* concepts. Even such concepts as *is a planet* or *is a moon of the Earth* are now understood to be functions. There is, however, something left out of the above discussion. To say that a function (or function-expression) is incomplete is also to say that the function (function-expression) can be completed—that it has values for arguments. A function-expression is completed by inserting an argument-expression in its argument place. The resulting expression designates the value of the function on the argument. When the function-expression is a predicate (a concept-expression) the resulting expression is a sentence. Sentences, then, must be object expressions and they must designate the values concepts have for arguments. What values do concepts have for their arguments? This is not a new question for Frege. Although the notion of function is not much developed in *Begriffsschrift,* Frege already identified concepts as functions there and indicated what their values are. The value of *is a planet* for the Earth as argument is, according to the views of *Begriffsschrift,* the conceptual content of the sentence, 'The Earth is a planet'. Sentences designate their conceptual contents. However, despite its initial plausibility, this answer will not work, for there is also a difficulty with Frege's use of the notion of conceptual content in *Begriffsschrift*.

The difficulty with taking sentences to designate conceptual contents can be made vivid if we consider Frege's *Begriffsschrift* discussion of identity statements. Frege tells us that identity statements are about signs. The content of the identity statement, '*A*=*B*', is that the signs flanking the identity sign have the same conceptual content, 'so that we can everywhere put *B* for *A* and conversely.' (*B* §8). But this will not quite do for the purposes of Frege's project. As we have seen, Frege wants definitions of the numbers that allow us to derive both basic truths of arithmetic and basic applications of arithmetic. Thus a definition of 1, for example, must enable us to show—if it has been established that there is a moon of the Earth, and that nothing else is a moon of the Earth—that 1 is identical to the number of moons of the Earth.

But do '1' and 'the number of moons of the Earth' have the same conceptual content? Conceptual content, as Frege characterizes it in *Begriffsschrift,* is content that has significance for inference. And '1' and 'the number of moons of the Earth' do not have the same significance for inference. To see why, consider the sort of inferences that can be made from the following two statements:

1 = the number of moons of the Earth

and

1 = 1.

The latter statement is an instance of a simple logical truth: that every object is identical to itself. It is neither particularly informative nor particularly useful. We can infer from it nothing more than what can be inferred from the primitive logical laws. The former statement, in contrast, is not an instance of a logical truth. We can infer from it that the Earth has a unique moon—something that does not follow from the primitive logical laws. Thus the signs '1' and 'the number of moons of the Earth' have different significance for inference, hence different conceptual content. The problem, it seems, is that sameness of conceptual content is too much to ask for when we want to establish identity. Indeed, Frege seems already to have abandoned this understanding of identity by the time he wrote *Foundations*, where he simply says that, if '*a*' is an object-expression, we must have a criterion for deciding in all cases whether *b* is the same as *a*.

It is evident from the above example that Frege is right to abandon the view that identity statements are about sameness of conceptual content. On Frege's new understanding, it seems, an identity statement is true just in case the same object is named on both sides of the identity sign—whether or not the expressions flanking the identity sign have the same conceptual content. This also has consequences for our understanding of the values certain sorts of functions provide for their arguments. For example, the value of the *author of* function, for the argument *Foundations*, is not the conceptual content of the expression 'the author of *Foundations*', but Frege. How can he provide a similar account of the values that concepts have for their arguments?

In order to answer this question, it will help to consider, first, what is involved in defining a function. Although Frege claims that definitions are identity statements, identity is a relation that can hold only between objects. A function sign cannot appear, on its own, on one side of the identity sign. This is evident since a function sign can never appear on its own—it must always be supplemented either by a variable or argument expression. But this is not a serious handicap for someone who wants to define a function. In general, the strategy for defining functions, even among those who are not engaged in Frege's project, is to use variables. Suppose, for example, one wants to define a function that takes a number, cubes it, and adds the original number to the result. It is usual to offer the following sort of definition: $f(x)=2.x^3 + x$. This is not exactly an identity statement, for since the expressions that appear on either side of the identity sign contain variables, they are not names of particular objects. Nonetheless, these expressions are object expressions, for they are complete. Although they do not pick out particular objects they do, to use Frege's locution, indefinitely indicate objects. The proposed function definition offered above is meant, not as an identity statement, but as a general statement. By substituting numerals for 'x' in the above definition we get names of particular objects on either side of the equal sign. The result is a number of identity statements, including:

$f(1)=2.1^3+1$
$f(2)=2.2^3+2$
$f(3)=2.3^3+3$, etc.

That is, our definition is a general statement from which we can infer what value $f$ has for each argument. Such a definition is not just a convenient way to define a function; it is, it seems, the only way to define a function. A function is identified by what values it has for its arguments.

Now suppose the function in question is a concept. How are we to define this sort of function? Frege writes in *Foundations* that a concept must either hold or not of each object. Just as a function can be defined by indicating what values it has for each argument, a concept can be defined by indicating, for each object, whether or not the concept holds of it. Or, in other words, what is required for a definition of a concept is something that indicates, for each object, whether it is *true or false* that the concept holds of the object. Consider, for example, *x is a square root of 1.*

If we replace *x* successively by –1, 0, 1, 2, we get:

–1 is a square root of 1,
0 is a square root of 1,
1 is a square root of 1,
2 is a square root of 1.

The first and third of these are true and the others are false. We can think of this concept as a function that gives us, for each object, either the answer 'true' or 'false'. Frege writes,

> I now say: 'the value of our function is a truth-value', and distinguish between the truth-values of what is true and what is false. I call the first, for short, the True; and the second, the False. (FC 13)

Supposing that the value of the function *is a square root of 1*, for the argument –1, is the True, then the sentence '–1 is a square root of 1' designates the True. All true sentences are, on the new view, to be regarded as standing for the same thing.

But, as Frege acknowledges, there are obvious objections to this view. The two sentences:

(1) The Evening Star is a planet with a shorter period of revolution than the Earth

and

(2) The Morning Star is a planet with a shorter period of revolution than the Earth

must stand for the same thing. For 'the Evening Star' and 'the Morning Star' name the same thing. Thus both of the above sentences pick out the value of the function *is a planet with a shorter period of revolution than the Earth* for the same argument. However, these sentences tell us different things or, to use a new locution that Frege introduces, express different thoughts. Frege now claims that a complete expression (not just a sentence) is associated both with an object, which he calls its *meaning* (we will discuss this odd choice of words in the next chapter) and with what it expresses, which he calls its *sense*. The expressions 'the Evening Star' and 'the Morning Star' pick out the same object but express different senses. This new notion of sense (thoughts are simply the senses of assertoric sentences) captures some of the notion of conceptual content that has been left out in Frege's new emphasis on what expressions stand for. But, as with conceptual content, sameness of sense is too much to require for identity. It is true that

The Morning Star=the Evening Star

although the expressions 'the Morning Star' and 'the Evening Star' express different senses. And it is, similarly, true that sentences (1) and (2) above name the same thing, although they express different senses. Although Frege devotes only a paragraph to the introduction of this new way of understanding the content associated with expressions, he also realizes that it is something that requires further explanation. He issues a promissory note in 'Function and Concept' and, indeed, returns to the issue in another of the papers published during this period, 'On Sense and Meaning', which we will discuss in our next chapter. The remainder of 'Function and Concept' is taken up with the introduction of the new version of his logic. We will examine, briefly, the modifications Frege makes to the workings of his logic.

The most important of these modifications have to do with Frege's notion of extensions of concepts. As we have seen, extensions of concepts play an important role in Frege's definitions of the numbers. Frege's solution to the Julius Caesar problem for the number one is to identify numbers as extensions of concepts. Since the notion of the extension of a concept, Frege claims, is a primitive logical notion, we can see, without elaborate justification, that Julius Caesar is not the extension of a concept and hence that the number one is not Julius Caesar. Frege also exploits the view that this is a primitive logical notion in 'Function and Concept'. All true sentences, he claims, designate the True and all false sentences designate the False. But what are the True and the False? And how do we know that neither one is Julius Caesar? Once again, Frege's answer is that these objects are extensions of concepts. But the notion of extension of concepts is new. In *Begriffsschrift* there is no discussion of extensions of concepts; there are no terms for extensions in the logical notation and no logical laws about extensions. If, as Frege claims, the notion of extension of concept is a primitive logical notion, then something is missing from the logic of *Begriffsschrift*.

What are extensions? We know, from the discussions of *Foundations*, that the extension of a concept is supposed to be a special sort of object associated with it. We also know one important characteristic of these associated objects: concepts that hold of exactly the same things have the same extension. But how are we to identify these objects? Frege claims, in 'Function and Concept', that concepts are simply functions. Just as a concept is a particular kind of function, an extension of a concept is a particular kind of value-range; a particular kind of object associated with a function. The notion of value-range is explained by turning, again, to the mathematical notion of function.

The writings of mathematicians, Frege claims, give us reason to think that there are special objects associated with functions. Functions are viewed as having associated curves. And functions that have the same value for each argument have the same associated curve. For example, for every value of $x$, $x^2-4x = x(x-4)$. Thus the functions $f(x)=x^2-4x$ and $g(x)=x(x-4)$ have the same value for each argument. Since these are functions of one variable, the existence of a single curve associated with $f$ and $g$ can be made vivid by representing the curve with a drawing on paper. The same drawing represents $f$ and $g$. Of course, since drawings on paper are two-dimensional, we cannot make such drawings for the curves associated with functions of more than two variables. Yet it still turns out to make sense for mathematical purposes to think in terms of multidimensional curves. Just as we can think of the curve associated with a function of two variables as occupying a region of (three-dimensional) space, we can think of the curve associated with a function of three variables as occupying a region of (four-dimensional) space-time. It is not difficult to generalize and think in terms of $n$-dimensional curves. On this way of thinking about functions, whenever functions have the same values on all arguments, there is an associated object that is the same.

The tendency of mathematicians to think of functions as associated with objects is also reflected, Frege claims, in the tendency of mathematicians to make statements in which function-expressions are used as if they were object-expressions. Thus, he says, we can also say:

> 'the value-range of the function $x(x-4)$ is equal to that of the function $x^2-4x$', and here we have an equality between value-ranges. (FC 9–10)

Given that—as Frege thinks is indicated by the writings of mathematicians—an understanding of the notion of function already involves the notion of the value-range of a function, this must be reflected in the logical notation. Moreover, Frege says in 'Function and Concept',

> The possibility of regarding the equality holding generally between values of functions as a particular equality, viz. an equality between value-ranges is, I think, indemonstrable; it must be taken to be a fundamental law of logic. (FC 10)

No such law appeared in *Begriffsschrift*. Indeed, no such law can be expressed in the logical notation offered in *Begriffsschrift*, for there is no way of forming the name of the value-range of a function using logical symbols alone. Thus if Frege is right that this is a fundamental logical law, something must be added to the logical language. Frege addresses the problem by introducing a symbol for a particular new logical function. This function takes functions as arguments and has, as its values, their value-ranges.

Frege's new logical function is very different from the functions we have seen so far. The functions we have seen so far take objects, not functions, as arguments. These familiar functions, because they take objects as arguments, are called first-level functions. The new function, because it takes functions (actually, first-level functions) as arguments, is called a second-level function. How are we to understand the notion of second-level function?

It will help to return to a more familiar logical notion, that of a quantifier. Consider the claim that there is at least one whale. This claim can be expressed using a quantifier:

$(\exists x)(x \text{ is a whale})$.

We can, as Frege argued in *Foundations*, regard this as expressing something about the concept *whale*. If we remove the expression 'is a whale' and replace it with another concept-expression, we get a sentence that expresses the same thing about a different concept. What is it that is stated about these concepts? We can express this by replacing 'is a whale' with a function variable, e.g.

$(\exists x)(fx)$.

This expression becomes a sentence (a name of a truth-value) whenever we replace '$f$' with a function-expression. Thus we can view the existential quantifier as signifying a second-level function—a function that takes first-level functions as arguments and has truth-values as its values.

The existential quantifier, of course, is not just a second-level function; it is a logical function—it can be represented in the logical notation using only expressions belonging to logic, expressions from the logical notation. In 'Function and Concept' Frege introduces a logical notation for a new second-level function that takes first-level functions to their value ranges. Using this notation the value range of a function, $f$, can be represented by the following expressions:

$\grave{\alpha}f(\alpha)$

and

$\grave{\varepsilon}f(\varepsilon)$.

Using this new notation, Frege can express the new logical axiom he needs: for any functions, say *f* and *g*, they have identical value-ranges if and only if they have the same value for every argument

$$(\grave{\varepsilon}f(\varepsilon)=\grave{\alpha}g(\alpha))=((x)(f(x)=g(x)))$$

[Note: There is one odd feature of the above expression: the use of the second identity sign as a translation of the English expression 'if and only if'. To see why this expresses the intended content, let us consider significance of a claim that, say, *P* if and only if *Q*. It is true that *P* if and only if *Q* just in case *P* and *Q* are either both true or both false—that is, either both name the True or both name the False, that is, $P = Q$.] Value-ranges of functions that are concepts are called extensions.

In addition to the changes described so far, there is another, minor change to the logical notation. The change amounts to obliterating the *Begriffsschrift* distinction between expressions for content that can become judgements (sentential expressions) from other expressions. To assert a judgement, in the first version of the logical language, one precedes the expression for its content with a content stroke to which a judgement stroke is attached. An assertion looks like this:

$\vdash A$.

The above can only be legitimately written, however, if '*A*' is, or is replaced by, a sentential expression. Other sorts of expressions (Frege uses 'house' as an example of this sort of expression) are prohibited from appearing, on their own, after the content stroke. It may seem that, in this respect, the workings of the logical language of *Begriffsschrift* mirror the workings of natural language. We can state, in natural language, that the Earth is a planet; but we cannot state that house. But Frege's point is not to mirror natural language. Although it would take us too far afield to see exactly how this works, the judgement stroke is needed in Frege's logical language as a technical device that makes it possible to express and evaluate certain inferences involving quantifiers.

The judgement stroke and content stroke are, of course, somewhat mysterious since they do not mirror anything that is explicitly stated in natural language. The new conception of sentential expressions as names for the True and the False enables Frege to eliminate the mystery from the content stroke, which he now calls 'the horizontal stroke'. To see how this works, it will help to consider the role played by the content stroke in *Begriffsschrift*. In everyday language assertion is indicated, not by a symbol or expression, but by context. If I were simply to say to you, in some context in which I clearly meant to be making an assertion, 'house', it would be reasonable for you to interpret this particular use of this term as having some sort of judgeable content—perhaps, 'there is a house'. But this works because in natural language we can use a word for different purposes in different contexts. The context provides the disambiguation. No such interpretation is available for the logical language, which is to be 'a system of symbols from which every ambiguity is banned' (CN 86). Thus Frege needs to include rules for the logical language that make it illegitimate to precede a non-sentential expression with a judgement stroke. He does this in *Begriffsschrift* by requiring that the judgement stroke be affixed to a content stroke and by prohibiting non-sentential expressions from appearing on their own after the content stroke.

In the new version of the logical language no such prohibition is necessary. The judgement stroke still needs to be attached to the horizontal stroke, but the horizontal stroke is now an expression for an ordinary function: *is identical to the True*. This function has as its value the True when the True is taken as argument. For any other argument, the value of this function is the False. Although this function can take as argument any object, even Julius Caesar or the number 4, this does not interfere with the proper use of the judgement stroke. For the horizontal is a concept-expression. Thus the judgement stroke can only be affixed to something that is, in effect, a sentential expression. Moreover, in the new logic the judgement stroke works, for sentential expressions prefixed by the horizontal stroke, exactly as it worked in the old logic. For example the sentence,

the Earth is a planet

is true (names the True) just in case

the Earth is a planet is identical to the True

that is, just in case

—— the Earth is a planet.

This may seem a minor, and very technical change. It is, however, from the point of view of many contemporary philosophers, part of a serious wrong turn.

What is the purported wrong turn? In natural language we distinguish between sentential and non-sentential expressions. This distinction was, it seems, recognized in *Begriffsschrift*, where the horizontal stroke can be followed only by a sentential expression. But the distinction is obliterated in the new version of the logic. Many think that this is a mistake. After all, sentences play a special role, not only in our use of natural language, but also in inference. Surely, one might suppose, this difference in our use of sentential and non-sentential expressions should be reflected in Frege's logical language. We need, it seems, a tripartite distinction of expressions into function-, sentence-, and object- expressions. In particular, sentences, as we use them in natural language, do not behave as names of objects. This is easy to see if we consider how we use particular natural language object-names and sentences. The string of words,

Julius Caesar is identical to the number one

is a perfectly good sentence (although not one anyone is likely to endorse). In contrast, the string of words,

The Earth is a planet is identical to the number one

is not a sentence at all. But, given the view of concepts introduced in 'Function and Concept', such a distinction can no longer be maintained. As we have seen, 'the Earth is a planet' is now regarded as designating an object, namely the True. Since each concept holds or not of each object, it holds or not, in particular, of this object. Presuming we accept this, it makes perfect sense to say, 'The Earth is a planet is identical to the number one'. Frege does not, of course, suggest that we introduce such sentences into natural language. But such sentences do appear in his logical notation.

Is this a difficulty with the new logical notation? It surely would be if Frege wanted the logical notation to represent the function of sentences in natural language. But the logical notation is meant to deviate from natural language in a number of ways. And there is no evidence that the tripartite distinction that is operative in our use of natural language has a role to play in Frege's logic, given the purposes it is meant to serve. Moreover, given his purposes, it is essential that sentential expressions be regarded as object-names. To see this, let us consider Frege's use of identity.

Although Frege distinguishes sentential from non-sentential expressions in *Begriffsschrift*, both sorts of expressions can appear flanking the identity sign. The reason is that identity, in the logic of *Begriffsschrift*, is simply sameness of conceptual content. A sentence, as much as any other expression, has conceptual content. Thus the identity sign is (and ought to be) usable to express the claim that two sentences have the same conceptual content. As long as identity is understood as sameness of conceptual content, the occurrence of sentences on either side of the identity sign does not force us to regard sentences as object names. That we are forced to do so is a consequence of Frege's new understanding of identity. On the new view of identity, expressions that can flank the identity sign must be object-names. One might think that, once Frege modified his view of identity, he should also have modified his rules. One might think that he should, in particular, have prohibited such identity statements as: 1 = the Earth is a planet. In order to see why Frege did not choose to modify his rules in this way, we need to consider the role played by functions and concepts in Frege's project.

One of Frege's tasks is to define concepts. A concept is a function and, consequently, is defined by indicating what its value is for each object. Suppose we use the expression '*E*' for the concept to be defined, and suppose we define it by saying that the property holds of the numbers that are divisible by 2. Our definition will look something like this: $E(x) = x$ is a number that is divisible by 2. The expressions on the each side of the identity sign become sentences when '*x*' is replaced with a name of a particular object. Were Frege to prohibit the appearance of a sentence on one side of the identity sign, he would be unable to offer this sort of definition of his concepts.

But it is not sufficient defence to say that Frege needs to view sentences as object-names in order to offer the definitions his project requires. After all, the criterion of adequacy for the logical language is not that it be usable to express Frege's definitions. Rather, the logical language is meant to be a tool for the correct evaluation of inference. The true test of the rules, in Frege's new logic, for the use of the identity sign and sentential expressions is the consequences these rules have for our evaluations of inferences. Frege turns to this issue in the next paper published during this period, 'On Sense and Meaning'.

# 6 *The reconception of the logic: II*

## 'On Sense and Meaning' and 'On Concept and Object'

Is identity a relation between objects or between names or signs of objects? This is the question with which Frege begins 'On Sense and Meaning'. As we have seen, Frege's answer will be that identity is a relation between objects. But there is at least an apparent difficulty with this answer. If identity is a relation between objects then we already know all there is to know about what is identical to what: each object is identical to itself and no other. Any true identity statement is simply a statement that a particular object is identical to itself and, consequently, obviously true. Yet not all true identity statements are obviously true. Frege writes,

> The discovery that the rising sun is not new every morning, but always the same, was one of the most fertile astronomical discoveries. (SM 25)

It was also an astronomical discovery that the morning star is identical to the evening star. How can we account for the fact that identity statements are informative?

Frege's *Begriffsschrift* answer was that identity is a relation between names—to say that $a$=b is to say that '$a$' and '$b$' have the same conceptual content. We have already seen why this answer will not work. Sameness of conceptual content is much too strong a requirement for identity. At least as we ordinarily understand identity, 1 is identical to the number of moons of the Earth yet '1' and 'the number of moons of the earth' have different conceptual content. Although he does not offer a general characterization of identity in *Foundations*, it is evident that Frege's understanding of identity has changed. This is manifested in the replacement of 'sameness of conceptual content', in his description of identity statements, with 'sameness of object designated'. His comments suggest that, at this point, he regards identity as a relation that holds between two names when they name the same thing. In 'On Sense and Meaning', however, Frege claims that such a view will not work. The use of a particular name to designate a particular object, he says, is arbitrary. If identity statements can express astronomical discoveries, then they cannot simply be statements about our arbitrary choices of which symbols to use for which objects. A new view is needed. He arrives at this new view by considering more carefully the content that is associated with object-expressions. In 'Function and Concept', Frege had already indicated that, in addition to the object it designates, an object-expression is connected with another sort of content: sense. He introduces the notion of sense again in 'On Sense and Meaning'—this time as a part of his answer to the question about identity.

The answer to the question requires the introduction of some terminology. Consider the expressions 'the morning star' and 'the evening star'. Both designate the same object, Venus. But each is associated with a different way of picking that object out or, to use Frege's expressions, a different mode of designation or mode of presentation. If we want to talk about an object we cannot do without a mode of presentation. An object expression, he claims, is associated both with an object that it designates (he calls this its *meaning*) and with a *sense*, which contains the mode of presentation of the object. A *proper name* is a term that 'has as its meaning a definite object . . . but not a concept or a relation' (SM 27). Finally, 'A proper name (word, sign, sign combination, expression) *expresses* its sense, *means* or *designates* its meaning' (SM 31).

Before we go on, it is worth paying some attention to Frege's choice of words in the passages quoted above. The words 'means' and 'meaning' in the above quotations are correct literal translations of the words *bedeutet* and *Bedeutung*, the words that Frege actually uses in these passages. Given Frege's use of these words, the meaning of a proper name is the object that it names. For example, the meaning of '2' is a particular object, a number; the meaning of 'Frege' is a particular person, Frege himself. Since this is a peculiar way to use these words, it will help to remember that, in the discussion that follows, the words 'means' and 'meaning' are to be understood in this way.

How does Frege's introduction of the notions of sense and meaning help to answer the question? Identity statements, on Frege's new view, *are* about the objects that are named. But, just as it is a mistake to identify the content of an object-expression with the object it designates, it is a mistake to identify the content of a statement (whether it is an identity statement or another sort of statement) as what it is about. Sentences, as well as the proper names that appear in them, express sense—sense that has as its parts the senses of its constituent expressions. Thus what is expressed by the sentence

the morning star = the morning star

is distinct from what is expressed by the sentence

the morning star = the evening star.

We can see, if we consider the senses expressed by these two sentences, why the former is trivially true but the latter is not. The sense of the former contains only one mode of presentation of an object while the sense of the latter contains both the mode of presentation associated with 'the morning star' and the (distinct) mode of presentation associated with 'the evening star'.

In addition to abandoning his talk of conceptual content, Frege also abandons his talk of significance for inference in favour of a new locution, 'value for knowledge'. The recognition of the truth of an instance of a simple logical law (for example, that the morning star = the morning star) has very little of this sort of value—it does not extend our knowledge. On the other hand, the recognition that the morning star = the evening star—a recognition that results from astronomical research—can extend our knowledge. For the purpose of acquiring knowledge, not only the meanings of our expressions (that is, the objects they designate) but also the sense of the expressions (that is, what they express) is important.

It does not seem at all unreasonable to say that we associate content (or sense) with object-expressions, in addition to an object designated. But we have not yet been given much of an explication of the notion of sense. It is a notion Frege never defines. He does, however, spend some time trying to get his readers to see what he means by 'sense'. One thing he makes clear is that, as he understands 'sense', the sense of an expression does not consist of the private ideas people associate with the expression. One person's idea, Frege says, is not that of another. But if two people speak a common language, there should be something common to their understanding of what is expressed by the expression—something other than that both use the expression to designate the same object. The differences between the meaning, sense, and ideas associated with an expression are illustrated, Frege thinks, by the following analogy,

> Somebody observes the Moon through a telescope. I compare the moon itself to the meaning; it is the object of the observation, mediated by the real image projected by the object glass in the interior of the telescope, and by the retinal image of the observer. The former I compare to the sense, the latter is like the idea or experience. The optical image in the telescope is indeed one-sided and dependent upon the standpoint of observation; but it is still objective, inasmuch as it can be used by several observers. (SM 30)

Ideas, like retinal images, cannot be shared. In contrast, a sense, like the image projected by the object glass, is something that can be shared by different people. The objectivity of something that is grasped by language speakers is evident, 'For one can hardly deny that mankind has a common store of thoughts which is transmitted from one generation to another' (SM 29). In light of this, Frege claims, we need have no scruples in speaking of *the* sense of an expression.

The suggestion of this last remark is that any two speakers of the same language associate the same sense with a particular expression. As Frege notes, however, natural languages contain object-expressions with which different speakers of the language associate different senses. Consider the name 'Aristotle'. While one person might understand this name as having the same sense as 'the pupil of Plato and teacher of Alexander the Great', another person might understand it as having the same sense as 'the teacher of Alexander the Great who was born in Stagira'. Although such variations of sense can be tolerated in natural language, Frege says, in a logically perfect language each proper name (or object-expression) must have a unique sense. Another requirement that a logically perfect language must meet but that a natural language need not meet is that each grammatically well-formed proper name must designate a unique object. Natural languages contain many proper names, for example 'Odysseus' or 'the celestial body most distant from the Earth', whose associated senses do not pick out any object at all.

Although each object-expression in a logically perfect language must express a unique sense and designate a unique object, there is no requirement that any language, even a logically perfect language, have a unique name for each object. Indeed, since a sense illuminates only a single aspect of an object, there may be a number of distinct senses and a number of distinct descriptions that pick out a particular object. It is an important feature of arithmetic, after all, that any number of numerical expressions ('6/3', '1+1', '13,578−13,576' etc.) can pick out the same number. This is one reason why identity statements are so important in mathematics. Frege's division of the content of an object-expression into the sense it expresses and the object it designates enables him to explain how these statements can have substantive content.

Frege's new understanding of the content of object-expressions seems to be all he needs to solve the puzzle about identity with which he begins 'On Sense and Meaning'. What is expressed by a sentence, and the significance that establishing its truth has for our knowledge, depends on its sense. Yet the introduction of the notions of sense and meaning occupies only the first few pages of his paper. It is not difficult to see why this is so. Nothing in the above discussion (and very little in the first few pages of the paper) indicates that the identity relation that he means to be discussing is anything other than what we understand as identity in our normal, everyday discussions. Indeed, while Frege indicates that natural language is defective, he also relies on his readers' understanding of how natural language works.

But, as we saw in the previous chapter, there is an important difference between our everyday notion of identity and the notion of identity that Frege exploits in his logic. In his logical language—unlike in natural language—sentential expressions can appear on either side of the identity sign. On the *Begriffsschrift* view of identity as sameness of conceptual content, the appearance of sentential expressions flanking the identity sign merely indicated that pairs of sentences, as well as pairs of everyday descriptions, can have the same conceptual content. But on the new account of identity statements, the fact that sentential expressions are permitted to flank the identity sign has new significance: sentential expressions must designate objects.

Frege might, of course, have avoided this consequence simply by deciding that his new account of identity warranted a change in the rules for the logical language—the introduction of a rule prohibiting the appearance of sentential expressions flanking the identity sign. But as we saw in the last chapter, Frege has a number of reasons for not introducing such a rule. One especially important reason is that concepts are functions. The result of filling in all argument spaces of function-expressions is an expression that designates an object. Because the result of filling argument spaces of a concept-expression is a sentence, sentences must designate objects. What objects do sentences designate? Frege's answer, in 'Function and Concept', is that sentences designate truth-values and, after introducing this rather strange view, he promises a further defence. Most of the rest of 'On Sense and Meaning' is devoted to this defence.

There are two parts to Frege's defence. The first part comprises a brief argument that sentences, in addition to expressing senses, have meanings. Some of our sentences contain names that have no meanings. Frege's example is, 'Odysseus was set ashore at Ithaca while sound asleep'. Supposing that 'Odysseus' does not name anyone, he says, the sentence in question is neither true nor false. Nonetheless, this sentence has a sense—Frege uses the term 'thought' for the sense expressed by an assertoric sentence. He writes, 'The thought remains the same whether "Odysseus" means something or not.' And in most circumstances we do not worry about whether the name 'Odysseus' names anybody. However, were we to take the sentence 'Odysseus was set ashore at Ithaca while sound asleep' to be either true or false, then it would be important that 'Odysseus' name someone. He continues,

> The fact that we concern ourselves at all about what is meant by part of the sentence indicates that we generally recognize and expect a meaning for the sentence itself. The thought loses value for us as soon as we recognize that the meaning of one of its parts is missing. We are therefore justified in not being satisfied with the sense of the sentence, and in inquiring also as to its meaning. (SM 33)

Frege says little more about why sentences must be regarded as the sort of things that have meanings. Most of his efforts are devoted to an argument that the meaning a sentence has is its truth-value.

Before we turn to this argument, it will be important to note one issue that does not arise in the discussions of 'On Sense and Meaning'. As we have seen, on Frege's view, linguistic expressions are connected with both a content that is expressed (a sense) and something else, a meaning. He also indicates that we care that the constituents of a sentence have meanings only when we expect the sentence to be either true or false. On this view it seems reasonable to say that most sentences are associated both with a sense that is expressed and with a truth-value. Moreover, if the meaning of an expression is simply whatever it is that contributes to the determination of truth-values of sentences in which the expression appears, then it would seem that the meaning of a sentence is its truth-value. However, it is also important to remember that to say that a sentence has a meaning is, on Frege's view, to say that it designates an object. From a contemporary point of view this is a peculiar claim. It is not just that our use of sentences in everyday contexts is very different from our use of names. It is also that, at least as most of us understand the term 'object', it is peculiar to think of truth-values as objects. Although one might expect Frege to devote some effort to making it plausible that sentences designate objects, he does not. One reason is that he discusses the notion of objecthood in another paper, 'On Concept and Object', written around the same time. But, as we shall see when we examine his discussion of objecthood, he is not in fact attempting to argue that, given our everyday notion of objecthood, we ought to regard truth-values as objects. Rather, Frege is introducing a new, logical notion of objecthood—a notion that has only limited connection to our everyday notion.

What, then, is the nature of Frege's argument in 'On Sense and Meaning' that sentences designate truth-values? Frege does not attempt to argue that this view fits our understanding of how sentences are used in everyday contexts. Rather, Frege wants his account to be true to the way sentences should be used in inference. The point seems to be, as he says about his definitions in the introduction to *Foundations* (*FA*, p. xi), not whether this view is natural but whether it goes to the root of the matter and is logically beyond criticism.

To decide whether it is logically beyond criticism to take the meaning of a sentence to be its truth-value, we need to decide whether the result can lead us astray in our evaluations of inference. How might we be led astray? Frege's new view of sentences as designating truth-values does not occasion any change in his logical laws and rules of inference. Most inferences will be evaluated as before. As we have seen, the change in Frege's view affects the truth-values of identity statements when sentences flank the identity sign. Thus if the new view introduces a problem, it is likely to result from uses of the laws governing identity. Throughout his career Frege maintains that, if $a$=b, then whatever is true of *a* is true of *b*. This is manifested, in both versions of the logic, by logical laws that license us to infer from *a*=*b*, and any statement about *a*—say *F(a)*—the same statement about *b*, that is, *F(b)*. Similarly, from *a*=b and *F(b)* logical laws license us to infer *F(a)*. All of this is encapsulated in a law of identity that appears in both versions of Frege's logic:

$$a=b \rightarrow (F(a) \rightarrow F(b)).$$

Another way of stating the significance this has for inferences is that, if *a*=b, then '*a*' and '*b*' are intersubstitutable in inferences.

This seems reasonable if '*a*' and '*b*' are everyday object-expressions. It is less obvious that we are entitled to such inferences if these expressions are sentences. In particular, it is less obvious given Frege's new view of identity. On the original view, to establish that *a*=*b* one needs to establish that '*a*' and '*b*' have the same conceptual content. It is evident that, if two sentences have the same conceptual content (the same significance for inference), they can be intersubstituted in inferences. But Frege now requires much less to establish *a*=*b*. For sentences, all that need be established is that they have the same truth-value. Supposing two sentences have the same truth-value, is this really sufficient to allow us to substitute one for the other in any inference? Most of 'On Sense and Meaning' is devoted to Frege's attempt to show that this is sufficient. If he succeeds, then he will have shown that his view that sentences mean or designate truth-values will not lead us astray in our evaluation of inference.

One of the keys to this attempt is a discussion that appears before Frege actually raises the issue of intersubstitutability of sentences. There is an apparent problem even with the intersubstitutability of everyday object-expressions. Consider the following argument:

Alice believes that the morning star is a body illuminated by the sun.
The morning star=the evening star.
Therefore, Alice believes that the evening star is a body illuminated by the sun.

Frege's law of identity allows us to infer from $a=b$ and from any statement about $a$, the same statement about $b$. The premisses of the above inference seem to include one that is a statement about the morning star and the assertion that the morning star is identical to the evening star. Thus, if we add the appropriate instance of the law of identity to the above argument, Frege's logic would seem to tell us that this is a valid argument.

But, if Frege's logic tells us this, something is wrong. For the second premiss is true. Yet, as he acknowledges, anybody who did not know that the evening star is the morning star might hold

> the morning star is a body illuminated by the sun

to be true but

> the evening star is a body illuminated by the sun

to be false (or vice versa). If so, and if Alice does not know that the morning star is the evening star, it could be that both premisses of the above argument are true but the conclusion is false. And if it could be that both premisses are true and the conclusion is false, the argument is not valid.

To see what has gone wrong here, let us consider the first premiss again:

> Alice believes that the morning star is a body illuminated by the sun.

What is this sentence about? It seems to be about Alice and what she believes. If Alice believes that the morning star is a body illuminated by the sun, what she believes is not a truth-value, but a thought. The contribution 'the morning star' makes to this thought (i.e. to what Alice believes) is its sense. If so, then in our attempt to determine whether this premiss is true or false, we must concern ourselves with the sense of the expression 'the morning star'. But, on Frege's view, it is not the senses of expressions that contribute to the truth (or falsehood) of sentences in which they appear. It is, rather, the meanings of the expressions (or, what they designate). Frege concludes that 'the morning star' does not have the same meaning in all contexts. In most contexts 'the morning star' has as its meaning a particular planet, Venus. But in the context of the premiss set off above, it has a different meaning. Instead of designating Venus in this context it designates what is ordinarily its sense.

This is not the only sort of sentence for which this problem arises. A similar problem arises if we were to say that Alice thinks (regrets, approves, hopes, fears, etc.) that the evening star is a body illuminated by the sun. Frege introduces the term *indirect* for this sort of context. Whenever an expression appears in an indirect context, he claims, its meaning is different from the meaning it has in usual (direct) contexts. In indirect contexts it designates its customary sense (that is, what it expresses in direct contexts). Thus, he says, the *indirect meaning* of a word is its customary sense. In indirect contexts the word also fails to have its customary sense (for the customary sense picks out the customary meaning). The sense the word has in indirect contexts, its indirect sense, is a sense that picks out its customary sense.

Frege's account of the meanings terms have in indirect contexts allows him to respond to the apparent counter-example we considered above. Let us consider the problematic argument again:

> Alice believes that the morning star is a body illuminated by the sun.
> The morning star=the evening star.
> Therefore, Alice believes that the evening star is a body illuminated by the sun.

Given that the morning star=the evening star, Frege's law of identity tells us that whatever can be truly asserted about the morning star can also be truly asserted about the evening star. It seemed that, in this argument, the first premiss is an assertion about the morning star and that the conclusion is the same assertion about the evening star. What Frege has argued is that the first premiss is not an assertion about the morning star and the conclusion is not an assertion about the evening star. Thus the law of identity does not provide us with licence to infer the conclusion.

But what about our earlier claim that the significance for inferences of Frege's law of identity is that if $a=b$, then '$a$' and '$b$' are intersubstitutable in inferences? The answer is that this characterization of the significance for inference works only for a logically perfect language. In a logically perfect language, as we noted earlier, each object-expression designates a unique object. In particular, what an object-expression designates is not dependent on context. In natural language, however, this is not so.

After considering the issue of whether, given that $a=b$, we are licensed to intersubstitute '$a$' and '$b$', when '$a$' and '$b$' are everyday object-expressions, he turns to the issue of whether such intersubstitution is licensed when these expressions are sentences. The remainder of 'On Sense and Meaning' is devoted to this issue. As he notes, it is impossible to survey all possibilities given by language. However, Frege surveys and discusses a large number of ways in which sentential expressions may appear as constituents of longer sentences. The survey does convince him to make one change in his view. Although he begins with the view that sentences designate truth-values, his more careful examination of the possibilities given by language shows us that sentences do not *always* designate truth-values. When he finishes this examination, he concludes a truth-value is 'the meaning of a sentence that has a thought as its sense' (SM 50). Although we will not examine the details of Frege's survey, it is important to see why he must modify his view. This is easiest to see by considering indirect contexts.

Just as we might originally have thought that the law of identity licenses the argument we discussed earlier, we might originally have thought that it licenses the following (invalid) argument:

Alice believes that the morning star is a body illuminated by the sun.
The morning star is a body illuminated by the sun = the evening star is a body illuminated by the sun.

> Therefore, Alice believes that the evening star is a body illuminated by the sun.

But, using Frege's analysis of indirect contexts, we can now see that the law of identity does not license this argument. Just as we saw earlier that 'the morning star' does not always designate its customary meaning, we now see that a sentence does not always designate its customary meaning. When a sentence appears in indirect speech it does not designate a truth-value but, rather, a thought. Both the first premiss and the conclusion are about Alice and her beliefs. The sentence 'the morning star is a body illuminated by the sun' has different meanings in the two premisses of the argument. It has its customary meaning in the second premiss, but in the first premiss it has its indirect meaning. Only in the second premiss does it express a thought and designate a truth-value. Since this expression has different meanings in these two premisses, the law of identity does not apply.

After his survey of the ways in which sentential expressions may appear as constituents of longer sentences, Frege offers an analysis of the circumstances in which a sentential constituent may not be replaced by another with the same truth-value. There are, he indicates, two sorts of cases. In some cases, as we have seen, the sentential constituent does not have its customary meaning (its truth-value). And, in these cases, the sentential constituent expresses, not its customary sense, but its indirect sense. Although Frege is not explicit about what exactly the indirect sense of a sentence is, it is evident, he thinks, that it is not a thought. There are also cases in which subordinate clauses play dual roles in a sentence. An example is the sentence:

> Bebel fancies that the return of Alsace-Lorraine would appease France's desire for revenge.

This sentence, Frege indicates, is a compressed way of expressing two thoughts:

> (1) Bebel believes that the return of Alsace-Lorraine would appease France's desire for revenge.

(2) the return of Alsace-Lorraine would not appease France's desire for revenge.

On this analysis, the sentence 'the return of Alsace-Lorraine would appease France's desire for revenge' occurs both directly (in (2)) and indirectly (in (1)). In order to play both roles, this sentence must designate both truth-value and thought. For this reason it will not do to replace this sentence with another sentence that shares only its truth-value. Frege concludes his discussion by saying,

It follows with sufficient probability from the foregoing that the cases where a subordinate clause is not replaceable by another of the same value cannot be brought in disproof of our view that the truth-value is the meaning of a sentence that has a thought as its sense. (SM 49–50)

In the final paragraph of 'On Sense and Meaning', he returns to his original question about identity and reminds the reader of how his introduction of the notion of sense solves the problem.

Before we leave our discussion of 'On Sense and Meaning', it is worth noting that Frege's discussion of apparent counter-examples to the law of identity shows us something about the significance of his logically perfect language. We can now see the importance of his requirement that there be no ambiguity in a logically perfect language. The apparent counter-examples that we examined above involve ambiguity. In each of the invalid arguments set off earlier, an expression or sentence appears with one meaning in one premiss and with a different meaning in another premiss. The identity law seemed to license these arguments because we presupposed, without noticing this presupposition, that the expression 'the morning star' has the same meaning throughout the argument. Frege's discussion drew our attention to the presupposition and showed us that it was false. Because the presupposition was false, the application of the law of identity was illegitimate. As long as our language is ambiguous, even given an inventory of the logical laws, it is not possible to tell whether the argument is valid simply by looking at its natural language expression. We must first assure ourselves that our tacit presuppositions (among which is that each expression has the same meaning throughout the argument) are correct. Frege's Begriffsschrift is meant to be a language in which the tacit attachment of presuppositions in thought is prohibited, a language in which no presupposition can sneak into an inference unnoticed. Since ambiguity is prohibited in Begriffsschrift, we need not presuppose that each term that appears in our Begriffsschrift expression of an inference has the same meaning throughout. It cannot be otherwise.

It is also worth noting that the arguments from 'On Sense and Meaning' constitute, in an important sense, a defence of Frege's notion of concept. For, as we have seen, the view he is defending in these discussions, the view that sentences that express thoughts designate truth-values, is a consequence of his view of concepts as functions. These arguments, however, do not handle all the apparent difficulties with Frege's notion of concept. In 'On Concept and Object', Frege responds to another.

In a paper published several years after *Foundations* appeared, Benno Kerry attacked Frege's claim that there is a fundamental distinction between concepts (functions) and objects. According to Kerry, even on Frege's view this is mistaken. For, Kerry argues, there is something—the concept *horse*—that is both a concept and an object. To see why, consider the statement:

> the concept *horse* is a concept easily attained.

This seems to be a true statement. It states that the concept *horse* falls under a concept, that of being a concept easily attained (that is, it is not difficult for us to understand what it is to be a horse). What falls under a concept must, as Frege understands 'object', be an object. Thus the concept *horse* is an object. But it is also, assuming the statement is true, a concept easily attained. Thus, the concept *horse* is a concept.

In 'On Concept and Object' Frege responds with an attempt to make his concept/object distinction clearer and to show, thereby, that the distinction between concept and object is absolute. Frege is careful to indicate, however, that this is not to say that Kerry's use of the terms 'concept' and 'object' is wrong. There are, Frege acknowledges, a number of ways to use these terms and his aim is simply to communicate and defend his own notions of concept and object—the notions that figure in his conception of his logic. In our examination of the arguments from 'On Concept and Object', it is important not to assume that these notions are either our everyday notions or contemporary notions. Although our contemporary first order logic comes from Frege's work, some of the details of his conception of logic are very different from contemporary conceptions of logic. Let us turn now to Frege's response to Kerry.

Frege does not deny that the concept *horse* is an object, but he does deny that it is a concept. A concept is predicative; that is, it is the meaning of a grammatical predicate. As we saw earlier, a concept is a kind of function; it is unsaturated or incomplete. An object, in contrast, cannot be the meaning of (what is designated by) a grammatical predicate—although it can be the meaning of a part of a grammatical predicate. Although Vesuvius cannot be the meaning of a grammatical predicate, it is the meaning of part of the grammatical predicate 'lives near the base of Vesuvius'. An object cannot be the meaning of a grammatical predicate (CO 193–4, 198). There is a simple criterion for distinguishing complex object-expressions from concept-expressions: object-expressions that contain articles contain definite, not indefinite articles. Thus 'the morning star' is an object-expression, while 'a planet' is a concept-expression. In the sentence

(1) The morning star is a planet

a concept, that of being a planet, is predicated of an object, the morning star.

One might think, however, that this does not quite work. For 'the morning star' means (or designates) the same thing as 'Venus'. Thus, if 'the morning star' is an object-expression, so is 'Venus'. And it may seem that in the sentence

(2) The morning star is Venus

Venus is predicated of the morning star. Thus Venus would be both an object-expression and concept-expression. But this is a mistake. The reason is that the word 'is' plays different roles in the two sentences set off above. In the first sentence the word 'is' is used merely as a copula—that is, a grammatical device to connect subject and predicate. But in the second, 'is' is used to express an identity statement. In this case, it is not inappropriate to regard 'is' as an abbreviation of 'is identical to': the thought can also be expressed by the sentence

The morning star is identical to Venus.

Our original sentence (1) can be regarded as stating that a particular object falls under a particular concept. But in (2) what is predicated of the morning star is not Venus itself, but the concept *is identical to Venus*. In the above version of the statement, Venus is the meaning of only part of the predicate. It is not (and can never be) the total meaning of a predicate. 'Venus' and 'the morning star' are both object-expressions.

Now let us return to the sentence with which we began, 'the concept *horse* is a concept easily attained'. Our simple criterion tells us that 'the concept *horse*' is an object-expression, for it begins with a definite article. But what of the other criterion? Can the concept *horse* be the meaning of a grammatical predicate? We certainly can say,

Venus is the concept horse.

But, as we saw in the discussion of 'Venus', this is not sufficient to show that the concept *horse* is the meaning of a grammatical predicate. As in the earlier case the word 'is' that appears in the above sentence appears as the 'is' of identity, not as the copula. That is, in this sentence the concept *horse* is not being predicated of Venus. On Frege's view, the expression 'the concept *horse*' cannot be used to designate a concept, and hence cannot be predicated of anything.

But this may seem to be the wrong way to go about trying to

predicate the concept *horse* of something. The problem, one might think, is that Frege is too concerned with grammatical role. And the solution is to adopt an alternative view on which the property of being predicative can hold of an expression independently of its grammatical structure. How might this work? Unlike 'the concept *horse*', the expression 'a horse' is predicative. In the sentence

(3) Venus is a horse

the word 'is' appears as a copula. If we do not insist that, because of their different grammatical roles, 'the concept *horse*' and 'a horse' must designate different sorts of things, we can say that the expression 'the concept *horse*' designates what is designated in (3) by 'a horse'.

This alternative view, however, runs afoul of Frege's understanding of identity. For Frege, if 'the concept *horse*' and 'a horse' designate the same thing, they must be intersubstitutable in inferences. That is, from

The concept *horse* is a concept easily attained

(which is a truth on the alternative view) we would be entitled to infer

(4) A horse is a concept easily attained.

But we are not. Indeed, due to the flexibility of natural language, both of the above are well-formed sentences. However, even if we accept the alternative view, the second does not follow from the first. For to begin a sentence with the expression 'a horse' is to say something about horses—either, depending on context, about all horses (as in, 'A horse is a mammal') or about a particular horse (as in, 'A horse is in my backyard'). Thus to say that a horse is a concept easily attained is to make a transparently false statement about horses.

We can also see from this that the problem is not typographical. It is not that, since we already use 'the concept *horse*' and 'a horse' for different things, we cannot make them designate the same thing. For if we attempt to choose a new, but similar, expression—for example, 'the concept designated by "a horse"'—as an object-name that means the same as 'a horse' we will run into exactly the same problem. In this case (4) should follow from the following sentence

The concept designated by 'a horse' is a concept easily attained.

But, again, it does not.

Indeed, we can now see that the problem is not grammatical at all. There is no difficulty in using the expression 'a horse' as the *grammatical* subject of a sentence. But the content of that sentence will be very different from the content of the sentence that results by replacing 'a horse' with any proper name, including 'the concept *horse*'. The distinction is not grammatical, but logical. Object-names play a different logical role from function-names. In the sentence,

The concept *horse* is a concept easily attained

'the concept *horse*' functions as an object-name. The sentence purports to tell us something about a particular thing, the concept *horse*; it purports to tell us that this particular thing is a concept easily attained. But if 'the concept *horse*' is replaced by 'a horse', the resulting sentence,

A horse is a concept easily attained

does not purport to tell us about a particular thing. Rather, it appears to be a general statement about horses—that they are concepts easily attained.

This difference would be evident were we to translate these two statements into a logical notation. For what needs to be substituted, for the object-expression in the first sentence, is not a concept-expression alone (to do that would result in an ill-formed expression) but, rather, a complex expression including, not only the concept-expression for 'a horse' but also a quantifier. In Frege's logically perfect language no confusion of the sort discussed above can arise. It is not that, in this language, we can formulate two propositions of the above sort and see that the one does not follow from the other. Rather, the rules for constructing expressions in Frege's logically perfect language prohibit us from substituting a concept-expression for an object-expression.

Using Frege's criteria, the concept *horse* is clearly an object, but not a concept. But it will not do simply to say that it is a consequence of Frege's criteria that the concept *horse* is not a concept. It is, as he admits, awkward to say that the concept *horse* is not a concept. After all, the city of Berlin is a city; the volcano Vesuvius is a volcano. And this is not the only awkwardness. As Frege says,

> In logical discussions one quite often needs to say something about a concept, and to express this in the form usual for such predications—viz. to make what is said about the concept into the content of the grammatical predicate. (CO 197)

Throughout his writings, for example, Frege uses the expression 'the concept number'. He means to be using this expression to talk about a concept but, as we have just seen, such an expression cannot be used to do this. He says,

> By a kind of necessity of language, my expressions, taken literally, sometimes miss my thought; I mention an object, when what I intend is a concept. I fully realize that in such cases I was relying upon a reader who would be ready to meet me half-way—who does not begrudge a pinch of salt. (CO 204)

Nor is the problem limited to such expressions as 'the concept number'.

In fact, any use Frege makes of the expressions 'concept' or 'function' will be problematic. To see why, we need only consider some of the consequences of the above discussion of the predicate 'is a concept easily attained'. As we saw, this can be predicated only of objects, not of concepts. One result is that, whenever it is predicated, the predication is false. That is, given Frege's understanding of this predicate, there is nothing that is a concept easily attained. The same is true of the predicate 'is a concept' or 'is a function'. As Frege himself says, somewhat later, the word 'concept' itself is, taken strictly, already defective, since the phrase 'is a concept' requires a proper name as grammatical subject and so, strictly speaking, it requires something contradictory. (*PW* 177–8/*NS* 192)

This appears to be a serious problem. It would seem, in particular, that there cannot be any content to Frege's various claims about the nature of concepts (functions).

One might expect that, once Frege recognized the extent of the problem, he would have felt called upon to revise his view that the notions of concept and function are the basic notions underlying his logic. Yet, while Frege explicitly, and repeatedly, acknowledged the defective nature of the predicates 'is a concept' and 'is a function', he never revised his view about the importance of these notions to his logic. Why?

Let us begin by considering exactly what Frege's statements about the nature of functions and concepts are meant to do. Kerry's objection, Frege says, is formulated as an objection to Frege's 'definition of "concept"' (CO 193). Frege responds that he has offered no definitions; that what is simple cannot be decomposed and what is logically simple cannot be defined.

On the introduction of a name for something logically simple, a definition is not possible; there is nothing for it but to lead the reader or hearer by means of hints, to understand the word as is intended. (CO 193)

Frege's statements about the nature of concepts are meant to be understood as hints. But it is not obvious that this will help. Is Frege hinting at a coherent notion of concept? Or is Frege simply making a mistake?

In order to answer these questions, we need to think more seriously about what the notion of concept is supposed to do for Frege. Defining, Frege says, must stop somewhere. The primitive terms on which a science depends cannot be defined and we must be satisfied with hints. Is 'concept' meant to be a primitive term of some science? If so there is a problem. For Frege's hints have been at least partly unsuccessful. Had they been entirely successful, Kerry would have understood and have offered no objections. But then it is not obvious that 'concept' *is* meant to be a primitive term of a science. Frege says that his use of the term 'concept' is a logical use. Presuming that it is meant to be a primitive term of some science, the science in question is logic. The truths of logic are to be expressed in the logical language, Begriffsschrift. Yet, there is no Begriffsschrift expression that corresponds to the natural language expression 'concept', as Frege uses it. This is not to say that we cannot express logical truths about concepts in Begriffsschrift. But the fact that something is a truth about concepts is indicated, not by predicating concepthood, but by the sort of symbols used. For example, in a logical language of the sort Frege introduced, the following sort of formula is used to say that each concept either holds or not of each object:

$$(F)(x)(F(x) \vee \sim F(x)).$$

In Begriffsschrift, as well as most contemporary versions of logical language, there are different categories of symbols and there are different rules for the use of symbols that fall under different categories. For example, the version of logical language used in the above formula, the symbols '$F$' and '$x$' are in different categories; the former is a concept-expression but the latter is not. Thus '$F(x)$' is a properly constructed expression but '$x(F)$' is not.

What is the significance of this for our understanding of Frege's comments about the nature of concepts and functions? Frege is more explicit in a later discussion of the predicative nature of concepts, where he says, once again, that he is not giving a definition. He continues,

> [F]or the decomposition into a saturated and an unsaturated part must be considered a logically primitive phenomenon which must simply be accepted and cannot be reduced to something simpler. I am well aware that expressions like 'saturated' and 'unsaturated' are metaphorical and only serve to indicate what is meant—whereby one must always count on the cooperative understanding of the reader. (*FG* 371–2)

Frege does not mean to introduce 'concept' as a name for something logically simple. Rather, his aim in using the terms 'concept' and 'object' is to get his readers to understand the functioning of Begriffsschrift and his logical regimentations.

This is not to say that all his uses of 'concept' have this character. Some of Frege's comments about concepts, even some of his defective comments, *are* meant to communicate literal truths. For example, he claims repeatedly, in *Foundations*, that concepts must either hold or not of each object. As we have just seen, this statement *can* be translated into Begriffsschrift. Indeed, it is a logical law. The defects of his actual statement are attributable to his use of natural language. But the logical truth, which Frege has attempted to state in natural language, can be given a literal and non-defective statement in Begriffsschrift—by a formula very like the one set off above.

It is important to understand the significance of the fact that Frege regards the primitive laws expressed in Begriffsschrift as the basic truths of his logic. The logical language plays a special role in the evaluation of inference. Once an inference is expressed in Begriffsschrift, it is possible to determine whether it depends on a necessary, but unstated, premiss. Because Frege's project is to determine the primitive truths that underlie the truths of arithmetic, his proofs of the truths of arithmetic must be expressed in the logical language—otherwise, there may be a hidden presupposition. The natural language statements about the nature of concepts and objects that do not ultimately surface as Begriffsschrift statements of logical laws are not properly regarded as part of Frege's explanation of what justifies the logical laws and the truths of arithmetic. Thus Frege's apparently paradoxical remarks about the nature of concepts and functions are no part of a logical theory that underlies the truths of arithmetic. These statements are merely hints designed to get Frege's readers to understand Begriffsschrift and its proofs. The true content of Frege's fundamental notions of concept, function, and object—the content that has significance for inference—is manifested in the logical laws and rules of

Begriffsschrift. Insofar as these laws and rules are correct, Frege's fundamental notions are coherent.

Nonetheless, many contemporary readers side with Kerry rather than with Frege on this issue. To see why, let us consider some of the consequences of Frege's response. Two of the central insights of *Foundations* are that ascriptions of number are assertions about concepts and that numbers are objects. Both of these statements appear to be substantive literal (and non-paradoxical) truths that constitute part of the foundation of Frege's theory of numbers. It is, however, a consequence of the view Frege describes in his response to Kerry that both must be relegated to the status of hints rather than that of literal truths. To see why, we must examine these statements individually.

Let us begin with the former statement. Since its expression contains the word 'concepts' it is, on this view, defective. When Frege ultimately translates statements of arithmetic into Begriffsschrift and proves them from logical laws, some of these logical laws are Begriffsschrift translations of his defective natural language sentences. But the claim that ascriptions of numbers are assertions about concepts is not among them. It is not, then, one of the truths that underlies the truths of arithmetic. This is not to say that, if we take Frege's response to Kerry seriously, we are precluded from offering any account of what Frege does with this statement. There is no obstacle to saying that Frege does use it to communicate. But the communication involved is not a matter of his stating, and our understanding, a literal truth. Rather, his success in using this statement to communicate is manifested in our recognizing the correct Begriffsschrift translations of ascriptions of number. One might suspect that it would be more appropriate to say that Frege really is stating a literal truth about correct Begriffsschrift translations of certain sorts of natural language statements. Moreover, one might think, there is no obstacle to expressing such statements in Begriffsschrift.

There is nothing really wrong with this reasoning. It is important to realize, however, that even if this is right—even if Frege means to be stating a literal truth about correct translations of natural language statements into Begriffsschrift —this literal truth is no part of the foundations of Frege's arithmetic. Frege intends to provide gapless proofs of the truths of arithmetic from the primitive truths on which they depend. These primitive truths are the foundations of arithmetic. If, among the primitive truths on which arithmetic depends, there are truths about the correct translations of natural language statements into Begriffsschrift, then Frege must provide gapless proofs of the truths of arithmetic from these truths about language. Gapless proofs, of course, can be provided only in Begriffsschrift. Thus, if he is to accomplish his aim, the truths that underlie arithmetic must be, not merely *expressible* in Begriffsschrift, but *expressed* in Begriffsschrift.

Frege's statement that numbers are objects is, from a contemporary point of view, even more important. Among contemporary philosophers there is a great deal of controversy about whether or not numbers are objects. Frege appears to have staked out and defended a position. Yet, although the term 'object' does not exhibit the problems associated with the terms 'function' and 'concept', it is a consequence of Frege's response to Kerry that there is no substantive issue here: nothing to debate or defend. Frege says, in his response to Kerry, that we can tell that the concept *horse* is an object because the expression 'the concept *horse*' can be used to fill in the blank space in '. . . is an object'. Thus, for Frege, any expression that can be used to fill in that space is an object expression. That is, anything of which objecthood can be predicated is an object. That is, everything is an object. And if everything is an object, it can hardly be informative to say that numbers are objects. This is not to deny that Frege uses this statement to communicate something. It is simply to say that, as with the statement about ascriptions of number, what is communicated is not a literal truth. The evidence that something has been communicated is manifested in our recognizing correct Begriffsschrift definitions of the numbers.

It is, thus, a consequence of Frege's response to Kerry that the claim that ascriptions of number are statements about concepts and the claim that numbers are objects are to be regarded as hints that get us to see that his definitions are correct. Neither is a part of his theory of numbers. Since at least one of these claims is the subject of contemporary debate and, indeed, both are claims for which Frege argues at length in *Foundations*, it may seem unreasonable that they should be regarded merely as hints. Many philosophers think that, were Frege to be confronted with this consequence, he would surely have retracted his response to Kerry. But the price to be paid for such a retraction would be very high. As we have seen, the role played by a 'concept' and 'object' in Frege's project is not trivial. Two of the central insights of *Foundations* are that ascriptions of number are statements about concepts and that numbers are objects. Both of these insights, as well as Frege's conception of his logic and logical language, would be undermined if Frege were to retract his response to Kerry.

The three papers of this period in Frege's career, 'Function and Concept', 'On Sense and Meaning' and 'On Concept and Object', introduced all of the modifications that he was to make to his language, Begriffsschrift, and his logical system. During this period he also completed his definitions of the natural numbers and some of the proofs of simple truths of arithmetic from these definitions and logical laws. In 1893, the year after 'On Concept and Object' was published, he published the first volume of his *Basic Laws of Arithmetic*. We will turn, next, to this work.

# 7 Basic Laws, *the contradiction and its aftermath*

As we have seen, in the years following the publication of *Foundations*, Frege reconceived his new logic. The changes are discussed informally in 'Function and Concept', 'On Sense and Meaning' and 'On Concept and Object'. His next publication, in 1893, was the first volume of *Basic Laws of Arithmetic*, in which he set out the new version of the logic and began the proofs that were to have brought the project to fruition. As Frege envisioned it, *Basic Laws* was to have several parts: a part in which the new version of the logic and its basic laws are set out, a part in which the natural numbers are defined and some basic laws governing them proved and, finally, a part in which the real numbers are defined and the foundations laid for assimilating analysis to logic. There is some evidence that Frege also envisioned defining the complex numbers.

The project Frege envisioned was never completed. Volume one of *Basic Laws* contains the first part of the project (the formal introduction of the new logic and its laws) and some of the second part (the definition of the natural numbers and proofs of some basic laws governing them). Ten years later, Frege published a second volume, which contains more proofs of the basic laws governing natural numbers and begins the investigation of the real numbers. Volume two of *Basic Laws* includes informal criticisms of other writers' views about the real numbers, sketches of Frege's proposed technique for defining the real numbers and, finally, a series of formal Begriffsschrift proofs that were to constitute the beginning of Frege's formal theory of real numbers. In 1902, when volume two was in press, he received a now-famous letter from Bertrand Russell showing that the second version of the logic was inconsistent.

Frege added an appendix to the second volume in which he discussed the contradiction and various strategies for avoiding it. Perhaps because he never found a satisfactory strategy for avoiding the contradiction, no third volume of *Basic Laws* was written.

Even if we disregard the contradiction, it may seem that, in the more than 500 pages that comprise the two published volumes of *Basic Laws*, Frege accomplished very little. By the end of volume two, Frege had not yet reached the point of defining the real numbers. Nor did he ever define even such basic operations as addition and multiplication for the natural numbers. Yet *Basic Laws* is an extraordinary achievement. Although the second version of the logic includes a new logical law (Basic Law V) which introduces an inconsistency into the logical system, this version of the logic is in other respects an important advance over the earlier version. The logical system, with Basic Law V removed, is a central contribution to the development of mathematical logic. But Frege's achievement is not limited to the construction of the logical system. There are substantial technical accomplishments in Frege's proofs that do not depend on Basic Law V. One example, which we discussed in Chapter 3, is Frege's use of laws about sequences to derive truths about the number sequence from logical laws. Contemporary readers continue to find that there is a great deal of both mathematical and philosophical interest to be learned from the investigation of the proofs and discussions of *Basic Laws*.

Although the discussions of *Basic Laws* are fascinating, they are also among the most detailed and technical of Frege's writings. Because the level of technical facility required to understand these proofs is so high, we will not be able to provide a comprehensive account of the achievements of *Basic Laws* in this short, non-technical introduction to Frege's writings. We can, however, examine the role that *Basic Laws* was meant to play in Frege's project and the significance Russell's contradiction had for this project.

Frege's project was to show that arithmetic is analytic. This required him to provide gapless proofs of the basic truths of arithmetic from premisses all of which are either definitions or basic logical laws; that is, premisses that require no evidence—in particular no evidence of the senses or intuition—for their support. Volume one of *Basic Laws* begins with an introduction of the second version of the logic and this introduction is followed by actual Begriffsschrift formulations of the definitions and proofs that were sketched in *Foundations*.

Frege claimed that *Basic Laws* contains 'the derivation of the simplest laws of numbers by logical means alone' (*BLA* §0). What are the simplest laws of numbers? It is natural to assume that these are the laws familiar to most of us from the study of elementary arithmetic, for example, the commutative and associative laws for addition. But while Frege indicated that he would like, ultimately, to prove such laws, the theorems proved in *Basic Laws* are more basic than these. Indeed, he did not even get so far as to define the addition function. Virtually the only theorem of *Basic Laws* that appears to be a familiar truth of elementary arithmetic is theorem 111—that 0 is not equal to 1.

Why is this? As we saw in earlier chapters, one of Frege's important discoveries is that many of the features of the natural numbers sequence can be proved from general propositions about sequences. Thus many of the theorems of *Basic Laws* are not about numbers at all but, rather, are general propositions about sequences and relations that will be needed to prove the requisite truths about the numbers. These theorems are used, in *Basic Laws*, to prove truths about the natural number sequence. Frege has already given examples of the sorts of propositions about the natural number sequence that must be proved: that 1 follows in the series of natural numbers directly after 0, and that each number can have only one immediate successor (*FA* §78). These appear in *Basic Laws* as theorems 110 and 71 respectively.

As Frege emphasized in *Foundations*, however, it will not do simply to prove truths about numbers. The numbers of systematic arithmetic 'should be adapted for use in every application made of number' (*FA* §19). It must be possible to prove, for example, that if 0 is the number belonging to some concept, then there is no object of which it holds (*FA* §75); that if 1 is the number belonging to a concept, then there is an object of which it holds (*FA* §78). These are theorems 94 and 113 of *Basic Laws*.

It is not difficult to see why Frege wanted to prove the theorems he proved. Moreover he clearly believed that our familiar truths of arithmetic could be proved from his theorems. But why did he not go on to prove these familiar truths? One might suspect it was because he believed that his proofs were sufficient—that, given these proofs, it is obvious that the familiar truths of arithmetic can be proved. But Frege required proof whenever proof is possible, and the familiar truths of arithmetic are not immediate consequences of the theorems of *Basic Laws*. Had he been interested simply in deriving truths about natural numbers from logical laws, he would surely have gone on to define such simple mathematical functions as addition and multiplication and prove some of the familiar laws that hold of these functions. He did not do this because his project was more grand.

It was Frege's aim to show that all mathematics, with the exception of Euclidean geometry, is analytic. The arithmetization of analysis made it reasonable to attempt to show this by providing logical foundations for arithmetic. These foundations for analysis require arithmetic for the complex numbers. The numbers with which Frege concerns himself in *Foundations* and the first volume of *Basic Laws*, however, are the natural numbers. Given that Frege has defined the natural numbers and proved truths about the natural number sequence, what should his next step be? It would be in line with mathematical practice to offer definitions of the addition and multiplication functions on the natural numbers immediately and to go on, later, to extend definition to wider domains such as the real and complex numbers. However, Frege objected to using this strategy for defining functions, which he labelled 'piecemeal definition', in the development of a systematic science.

Systematic science must be based on a firm foundation. Once introduced, the meaning of a sign cannot later be changed. Once proved, a theorem cannot later be shown to be false. But to use the piecemeal definition strategy is to change the meanings of function signs whenever the domain is extended and, consequently, to change the theorems. Suppose, for example, the expression 'square root of 9' were defined first on the domain of positive integers and later extended to the domain of positive and negative integers. The original definition would allow us to prove that there is only one square root of 9. But, once the domain is extended to negative integers, this theorem no longer holds—there are two square roots of 9, 3 and –3. He writes:

In this way we never have really firm ground underfoot. If we have no final definitions we likewise have no final theorems. We never emerge from incompleteness and vagueness. (*BLA*, vol. ii §61)

For systematic science, the proper strategy is to define such familiar functions as addition and multiplication only after the most general concept of number (that is, complex number) is defined.

Frege turned his attention to the real numbers in volume two of *Basic Laws*. He criticized the views of others and suggested his own strategy for defining the real numbers. But although the proofs of volume two of *Basic Laws* do some of the preliminary work, the real numbers are not actually defined. It is a more complicated task to define the real numbers than it is to define the natural numbers, and it was a task that Frege meant to leave for the next volume. When volume two was in press, however, a new obstacle presented itself. Russell's letter arrived with a demonstration that the new version of Frege's logic was inconsistent. The problem Russell had found was disastrous. Without the modification of Frege's logic that led to the inconsistency, the introduction of value-ranges, there was no way to provide purely logical definitions of the numbers.

Why are value-ranges so important? Let us recall Frege's requirements for his definitions. A definition of, for example, the number one must specify exactly what the number one is. The proofs of truths about the number one will be purely logical proofs only if this definition is expressible in purely logical terms. Thus Frege needed to find objects that could be identified in purely logical terms. Moreover, these identifications needed to determine, among other things, whether or not the number one is Julius Caesar. His strategy was to exploit the traditional logical notion of the extension of a concept. An extension is supposed to be an object associated with a concept that has as its members precisely those things of which the concept holds. Assuming we understand what extensions of concepts are, it is evident that Julius Caesar is not the extension of a concept. Frege was able to show how it was possible to define the number one as an extension of a particular concept and to use this definition, along with logical laws, to prove truths about the number one. Although he claimed, in *Foundations*, to attach no importance to bringing in extensions, Frege soon realized that it was essential to bring in extensions. In the preface to volume one of *Basic Laws* he says that we cannot get along without them (*BLA*, p. x).

But what are extensions of concepts? In Frege's new logic, the basic notion is that of function, not concept. Most familiar mathematical functions (e.g. $x+2$) have numbers as values. But, as we saw in Chapter 5, Frege's notion of function is more general. A concept is a particular sort of function, a function all of whose values are truth-values. The concept *is identical to 2*, for example, has the True as value when its argument is 2 and has the False as value for all other arguments. The traditional logical view that there is an object, an extension, associated with each concept is now a special case of the view that there is an object associated with each function. Frege called these objects value-ranges. Extensions of concepts are value-ranges of concepts. But if, as this suggests, the laws that hold of value-ranges are logical laws, there is something wrong with the first version of Begriffsschrift. For there is no way to express these laws. Frege rectified this in the second version.

As we saw in Chapter 5, the strategy for introducing logical symbols for value-ranges is described, along with the other changes to the first version of Begriffsschrift, in 'Function and Concept'. Frege writes,

The possibility of regarding the equality holding generally between values of functions as a particular equality, viz. an equality between value-ranges is, I think, indemonstrable; it must be taken to be a fundamental law of logic. (*FC* 10)

And Frege introduces a new Begriffsschrift symbol for a second-level function that takes, as arguments, first-level functions and gives, as values, their value ranges. Using this symbol, we can name the value-range of any function: if $f$ is a function, then $\grave{\alpha}f(\alpha)$ is its value-range.

The introduction of the means for forming a name for the value-range of a function carries with it an implicit assumption: that any function that can be named *has* a value-range. Otherwise it would be possible to form Begriffsschrift object-names that did not name anything—something that Frege claimed, in 'On Sense and Meaning', cannot occur in any logically perfect language (SM 41). Thus one might expect the introduction of the symbol to be preceded by a proof that the requisite value-ranges exist. In 'Function and Concept' there is no sketch of such a proof. Frege simply drew his readers' attention to features of the writings of mathematicians that, he said, indicate that an understanding of the notion of function already involves the notion of the value-range of a function.

How, then, is the symbol introduced in *Basic Laws*? Since it is to be a primitive Begriffsschrift symbol, it cannot be defined in Begriffsschrift. Moreover, this second-level function is something logically simple—and what is logically simple, as Frege first indicated in 'On Concept and Object' and repeated in §0 of *Basic Laws*, is not properly definable. There can be, then, no definition in everyday language either. In such cases, he claims, the only option is to use hints to lead the reader to understand the symbol as intended.

It is important not to mistake the significance of this appeal to hints. One might initially think that this is a desperate move, indicative of a difficulty with the notion of value-range. Yet Frege did not introduce the appeal to hints to deal with this difficulty. Even in his earliest descriptions of his project, Frege argued that defining must stop somewhere; that at some point in our analysis we reach terms that are not definable. Nonetheless, there must be some way for the person who introduces terms for what is logically simple to communicate their meaning. In §0 of *Basic Laws* Frege claimed that this can be done only by means of hints. Moreover, he saw nothing problematic in his hints at the meaning of most of the primitive terms of Begriffsschrift. Yet, while the appeal to hints was not a desperate move designed to handle worries about the legitimacy of assuming that each function has a value-range, there is no question that Frege had such worries.

The introduction of the new symbol in §9 of *Basic Laws* is preceded by an introduction of the notion of value-range via an account, in everyday language, of the new logical law, Basic Law V. The claim that functions have the same values for each argument, he says, can be transformed into the claim that these functions have the same value-ranges and vice versa. Moreover, this must be regarded as a law of logic. And why should Frege's readers believe this? As in 'Function and Concept', they are referred to a practice with which he presumes they are familiar—that of talking about extensions of concepts. Such talk, he claims, implicitly relies on this law, as does the Leibniz-Boole calculus of logic. He also adds another sort of defence that is both surprising and unconvincing. He reminds his readers that he had, in *Foundations*,

> defined a Number as the extension of a concept, and indicated then that negative, irrational, in short all numbers were to be defined as extensions of concepts. (*BLA* §9)

He then offers, without discussion or apology, an apparently circular justification—the introduction of value-ranges is indispensable for these definitions. These remarks are immediately followed by the introduction of the new symbol.

His discomfort with this introduction is apparent. For the very next section is entitled 'the value-range of a function more exactly specified'. And even this more exact specification did not satisfy him. An elaborate argument that his second-level value-range symbol designates something appears in a later section (31). Although this may not seem surprising in a section whose explicit aim is to show that all the simple names do designate something, the argument devoted to the value-range symbol is far more elaborate than those devoted to his other symbols. Most of the latter arguments are fairly trivial—some amount simply to the claim that the conclusion follows immediately from his earlier explanations.

Did Frege regard the argument in section 31 as a proof that the value-range function symbol designates something? This question is not easily answered. Section 31 contains some of the most difficult and obscure writing in Frege's corpus. Contemporary writers are divided both about how the technical details of the arguments are supposed to work and about what goes wrong and why. Contemporary writers are also divided about how Frege understood what he was doing. If he meant the introductions of his primitive terms to be taken merely as hints—which, as he mentions in a later work, cannot be guaranteed to work—why should he have devoted a section to a series of arguments (apparently proofs) whose aim is to show that his primitive terms do designate something? Surely if such proofs are possible, then it is possible to introduce the value-range symbol by some means more secure than a hint. Yet, as we have seen, Frege's introduction of this symbol does look like a hint—there simply is no precise and explicit explanation of what this symbol is to mean. He certainly offers no explanation that seems usable as the premiss of a proof that this symbol designates something.

Moreover, if we attempt to read section 31 as setting out such a proof, it is difficult to identify its premisses and the structure of the argument. This is especially strange given the aims of *Basic Laws*. If the arguments of section 31 are meant as part of the ultimate justification of the truths of arithmetic, why did Frege not attempt to make sure they are gapless by expressing them in Begriffsschrift? And, if they are not part of the ultimate justification, why should proof be required? Finally, if he believed he had a gapless proof that his new symbol designates something, why did he continue to have reservations about it? For it is evident that he did continue to have reservations. After indicating that he expects the careful reader to be convinced by virtually everything in his book, he says in the preface,

> A dispute can arise, so far as I can see, only with regard to my basic law concerning courses-of-values (V), which logicians perhaps have not yet expressly enunciated, and yet is what people have in mind, for example, where they speak of the extensions of concepts. I hold that it is a law of pure logic. In any event the place is pointed out where the decision must be made. (*BLA*, p. vii)

Frege's reservations were warranted. The decision to accept Basic Law V, along with Basic Law VI—the other law in which the new symbol appears—resulted in an inconsistent logical system.

The inconsistency will be easier to understand if we focus on a particular consequence of these two laws. On the traditional view to which Frege appealed, an object of which a particular concept holds is said to be in its extension. A concept's extension is viewed as having, as its members, all objects of which the concept holds. This is easily proved using Basic Law V and Basic Law VI (which introduces a definite description operator and tells us, in effect, that *a* is the unique member of the extension of the concept *is identical to a*). Theorem 1 states that an object is a member of the extension of a concept just in case the concept holds of it. Using the symbol 'ε' for 'is a member of', we can express theorem 1 in symbols as follows:

$$(x)\,(x \;\varepsilon\; \grave{\alpha}f(\alpha) \leftrightarrow f(x))$$

The contradiction is easily proved from theorem 1.

Consider a particular concept, the concept *fixed star*. Assuming that this is a legitimate concept, each object either is, or is not, a fixed star. Those objects that are fixed stars are members of its extension. The extension itself is an object and we can ask whether or not this object is a member of the extension of *fixed star*. Since extensions of concepts are not fixed stars, the extension of the concept *fixed star* is not a member of itself. Indeed, most extensions are, presumably, not members of themselves. That is, the concept of not having itself as a member is a perfectly good concept—a concept that seems to hold of most extensions and, indeed, of most other objects. Let '*Ax*' abbreviate '*x* is not a member of itself' or, to be more explicit:

*x* is not a member of *x*.

Now consider $\grave{\alpha}A(\alpha)$, the value-range of *A*—in other words, the extension consisting of precisely those objects which are not members of themselves. Let us use '*b*' to abbreviate this expression. By theorem 1,

for any *x*, *x* is a member of *b* if and only if *Ax*.

And this is just to say that:

for any *x*, *x* is a member of *b* if and only if *x* is not a member of *x*.

This, of course, is a universal generalization. In Frege's logic (both early and late) as well as in contemporary logic, every instance of the universal generalization is derivable from the universal generalization. In this particular case, we have a universal generalization that tells us that, no matter what *x* is,

*x* is a member of *b* if and only if *x* is not a member of *x*.

If the above holds of every object then, in particular, it holds of the object *b*. We can derive the following:

(*) *b* is a member of *b* if and only if *b* is not a member of *b*.

This is the problem to which Russell drew Frege's attention. Although (*) does not have the form of a classical contradiction ('*P* &~*P*'), it implies something that does: that *b* is a member of *b* and *b* is not a member of *b*. To see this, suppose that *b* is a member of *b*. Then by (*) it follows that *b* is not a member of *b*. We can conclude that *b* is not a member of *b*. But, given the *b* is not a member of *b*, it follows from (*) that *b* is a member of *b*. Thus we can also conclude that *b* is a member of *b*. Hence we can conclude: *b* is not a member of *b* and *b* is not a member of *b*.

Russell's letter was dated 16 June 1902. Only a few days later, on 22 June, Frege responded, acknowledging that this discovery had 'rocked the ground on which I meant to build arithmetic' (Frege to Russell 22 June 1902). Although volume two of *Basic Laws* was already in press, Frege was able to write an appendix with an explanation of the contradiction and a discussion of various strategies for avoiding it. Frege introduced a new law (V′) to replace law V, a replacement that, he thought at the time, would both eliminate the inconsistency and allow him to prove his original results.

In the years immediately following the publication of *Basic Laws*, Frege published an extensive criticism of another project, the attempt by the mathematician David Hilbert (1862–1943) to provide foundations for geometry. One notable feature of this work is Frege's concentration on the issue that got him into trouble—the introduction of scientific terms. He attempts to articulate what is required to introduce a scientific term—either by definition or, in the case of simple terms, by hints (or, as he now calls them, 'elucidations'). During this period Frege seems to have been convinced that he would be able to fix the system.

At some point, however, Frege came to the conclusion that the project he had set himself could not be accomplished—that it could not be shown that the truths of arithmetic are analytic. In a note dated August 1906 and entitled 'what may I regard as the result of my work?', there is no mention of a strategy for showing that truths of arithmetic are analytic. The note, in its entirety is,

It is almost all tied up with the concept-script. a concept construed as a function. a relation as a function of two arguments. the extension of a concept or class is not the primary thing for me. unsaturatedness both in the case of concept and functions. the true nature of concept and function recognized.

. . . strictly I should have begun by mentioning the judgment-stroke, the dissociation of assertoric force from the predicate . . .

hypothetical mode of sense composition . . .

generality . . .

sense and meaning (Bedeutung) . . . (PW p. 184/NS p. 200)

Much of Frege's work in the remaining years of his career was devoted to writing a non-technical introduction to his logic. His unpublished writings contain a number of outlines and beginnings of such a work. A first part of the intended introduction appears in his last publication, a series of three papers entitled *Logical Investigations*. The three papers were entitled 'The Thought', 'Negation', and 'Compound Thoughts'. They begin with a discussion of the notion of truth and the sense in which, on Frege's view, the laws of logic are the laws of truth. In 'Negation' and 'Compound Thoughts', he goes on to discuss the truth-functional connectives. But Frege never completed the introduction. He never got to the discussion of generality and to the introduction of his logical notation.

Most readers today regard 'The Thought' as the most interesting part of *Logical Investigations*. This essay is particularly renowned for its assertion that there is, in addition to the external world of physical objects and the internal world of ideas, a third realm of non-spatio-temporal objective objects, among which are thoughts. The notions of thought and truth are, Frege claims, simple and indefinable. The meaning of the word 'true' is, he says, spelled out in the laws of truth (T 59). He characterizes a thought as something for which the question of truth can arise (T 60).

From this information one might infer that thoughts are primitive objects and truth a primitive property that holds or not of thoughts. Yet, after discussing the sense in which thoughts can be true, he writes,

> May we not be dealing here with something which cannot be called a property in the ordinary sense at all? In spite of this doubt I will begin by expressing myself in accordance with more ordinary usage, as if truth were a property, until some more appropriate way of speaking is found. (T 61–2)

Of course, in *Basic Laws*, Frege did introduce another way of speaking about truth. In the discursive natural language discussions that appear in *Basic Laws*, Frege typically does not use the expression 'is true' but, rather, 'is the True'—a natural language expression that translates his horizontal function. But this expression is not a truth predicate. The True is a particular object that, as Frege said in 'On Sense and Meaning', is recognized 'if only implicitly, by everybody who judges something to be true' (SM 34). And neither sentences nor thoughts have the property of being the True. Although Frege mentions the True and the False in some unpublished notes written about the same time as 'The Thought', there is no way to tell whether Frege meant to introduce a more appropriate way of speaking along these lines. For there is no mention of the True in *Logical Investigations*. The views about truth and thoughts expressed in 'The Thought' continue to be a matter of fascination and controversy to philosophers today.

Frege's attempt to provide a non-technical introduction to his logic was not the only work that occupied him late in his career. For, even after the collapse of his original project, the subject, as Frege said in the preface to *Basic Laws*, would not let him go. Notes from a course that he taught in 1914 contain a detailed discussion of the notion of a systematic science and the strategy of providing a foundation for a science by systematizing it. Correspondence from as late as 1918 indicates that he still thought that there must be some way to clarify what we mean by the word 'number' by replacing the notion of extension with some sort of notion of class. Although he ultimately gave up on this, he continued to work on foundations for arithmetic. Frege came to believe that the geometrical source of knowledge is also the source of our knowledge of the truths of arithmetic and he began to work on strategies for providing a geometrical foundation for arithmetic. In a fragmentary manuscript entitled 'A new Attempt at a Foundation for Arithmetic' that was written in the last years of his life, he discussed a strategy for defining the numbers using the basic notions of *line* and *point*. Some of his last extant writing is part of a correspondence in April 1925 about publishing an account of this new strategy. The editor to whom he wrote expressed interest in Frege's ideas and asked

Frege if he would expand his manuscript and resubmit it for publication as a monograph. Unfortunately, we will never know how Frege would have worked out his attempt at a new foundation from arithmetic. He was already ill by the time he corresponded with the editor. Frege died on 26 July 1925.

# 8 *Frege's influence on recent philosophy*

The project to which Frege devoted nearly his entire career—his attempt to show that the truths of arithmetic are analytic—ended in decisive failure. Yet the story of Frege's philosophical work is not the story of a failure. His writings are immensely influential today. His logic was unquestionably an important advance both for mathematicians and for philosophers. Frege's critique of a venerable philosophical argument, the ontological argument for the existence of God, is one of the earliest demonstrations of how his logic can be used, as it now is routinely, to clarify arguments.

The ontological argument begins with a definition: God is the most perfect being that can be conceived. We can make a number of inferences about God's nature using this definition. For example, an omniscient being would be, presumably, more perfect than a being who is not omniscient. If so, we can see from the definition that God must be omniscient. Similarly, going down the list of properties that God might or might not have, we can see that God must be omnipotent, omni-benevolent, etc. What about the property of existence? An omniscient, omnipotent, omni-benevolent, etc. being that does not exist would be, presumably, less perfect than an omniscient, omnipotent, omni-benevolent being who did exist. If so, a being that does not exist is *not* the most perfect being that can be conceived. Existence is one of the properties of the most perfect being that can be conceived. Hence God exists.

But is existence a property? The evidence that it is comes from the grammar of natural language. We can say 'God is omniscient' and, similarly, 'God exists'. Both statements appear to predicate a property—omniscience in the first, and existence in the second—of God. However, one of the morals of Frege's new logic is that grammar is not a good guide to the analysis of a statement. And, given the analysis introduced by Frege's logical regimentation, existence is not a property.

On Frege's logical analysis, the predicate 'exists' is actually a natural language expression for a quantifier. To say that a horse exists is not to predicate existence of a particular horse but rather, to say that there exists an $x$ such that $x$ is a horse (in symbols, '$(\exists x)Hx$'). Or (using inexact natural language) to say this is to predicate something of the concept *horse* (i.e., that something falls under it). How does this analysis bear on 'God exists'? The ontological argument begins with a definition of God—as the most perfect being that can be conceived. Thus to say that God exists is to say that there exists something that is the most perfect being that can be conceived. On this analysis, 'exists' is used to say something about the concept *most perfect being that can be conceived*; existence statements are about concepts. There is no property of existence that can be predicated of an object. And so the argument does not go through.

Moreover, there is another problem. In fiction we can use names that do not name anyone and descriptions that do not describe anything. However, in other contexts, we are entitled to use expressions of the form 'the *A*' only if we can first establish the existence of an *A*—otherwise the description fails to pick anything out. Hence the legitimacy of using the expression 'the most perfect being that can be conceived' to define God depends on our being able, prior to the introduction of the definition, to establish that there *is* a most perfect being that can be conceived. Thus the argument is circular. We are not entitled to introduce the premiss containing the definition of God unless we have already established the conclusion—that God exists.

This is but one example of how Frege's new logical regimentation can be used in the evaluation of a philosophical argument—albeit an interesting one in so far as it illustrates the importance of his divergence from the traditional view that grammatical structure is a guide to logical structure. But there are many others.

As we have seen, Frege believed that he could solve a philosophical problem about the nature of the truths of arithmetic by answering the question, what is the number one? On the one hand, since they have applications throughout the widest domain of all, the truths of arithmetic seem to be logical (analytic) truths. On the other hand, because they seem not to be general truths but truths about particular objects, the numbers, they seem *not* to be logical truths. Frege's strategy for solving the problem was to introduce definitions, using purely logical terms, that could replace numerals in all contexts. To show that his definitions would do the job, he undertook to show that the truths about numbers could be derived from these definitions and logical laws. The philosophical question that Frege wanted to answer appears to have nothing in particular to do with language or meaning. Yet he answered the question by engaging in a linguistic investigation—an analysis of how certain problematic symbols (numerals) are used and a demonstration that we can dispense with these symbols by defining them from non-problematic terms.

The use of this strategy marks Frege as one of the first (and perhaps the very first) to take the so-called 'linguistic turn' that is characteristic of analytic philosophy, the dominant school today in Anglo-American philosophy. W. V. O. Quine (b. 1908), who calls it 'analysis' or 'explication' offers a particularly good characterization. Quine says that analysis is called for when we have 'an expression or form of expression that is somehow troublesome'. That is,

> It behaves partly like a term but not enough so, or it is vague in ways that bother us, or it puts kinks in a theory or encourages one or another confusion. But it also serves certain purposes that are not to be abandoned. Then we find a way of accomplishing those same purposes through other channels, using other and less troublesome forms of expression. (*Word and Object* (MIT Press, 1960), 258)

Although Frege once claimed that empiricists would least appreciate his results (*FA*, p. xi), his method of analysis has been particularly favoured by empiricists. Suspicious of such abstract objects as numbers, classes, and propositions, which do not seem accessible by sense experience, empiricists attempt to show that we can define them away in terms of unproblematic objects.

Many contemporary philosophers have adopted another part of Frege's perspective as well. Frege's aim was not simply to banish troublesome terms but also to reduce the number of undefinable terms. He believed that, in order to understand the nature of a discipline, we must reduced the entire discipline to the fewest primitive truths using the fewest simple indefinable notions. It is this procedure that identifies, to use Frege's locution, what a theory's ultimate building blocks are or, in more contemporary language, what a theory's ontological commitments are. In all these activities, a contemporary version of Frege's logical regimentation has been crucial. Ontological commitments of a statement are routinely measured by the terms required to express the statement in logical notation.

Frege's logic has been important for mathematicians and computer scientists as well. The field of mathematical logic has its origin in Frege's new logic. Although the logical system set out in *Basic Laws* was inconsistent, the inconsistency is easily eliminated by omitting Basic Law V. The resulting logical system has more power than those it replaced. It has all the power necessary for it to be used to identify, as logically valid, all inferences that are regarded today as logically valid. Moreover, Frege's logic is formal in several important respects. It is a mechanical task to determine, for any string of symbols in the logical language, whether it is a well-formed name or sentence of the language. Moreover, it is a mechanical task to determine whether a sentence is an axiom and whether or not a sentence follows immediately by one of the rules of inference from other sentences. Thus, although there is no mechanical procedure for using the logic to determine whether or not an argument is valid, there is a mechanical procedure for evaluating a purported gapless proof of the argument in the formal language. The formal nature of Frege's logic made it possible to regard the logical system itself as a mathematical entity.

Mathematical logic, however, also involves an important departure from Frege's conception of logic—a departure that has become an important part of the contemporary conception of logic. Frege wrote,

> The questions why and with what right we acknowledge a law of logic to be true, logic can answer only by reducing it to another law of logic. Where that is not possible, logic can give no answer. (*BLA*, p. xvii)

He regarded Begriffsschrift as a language that expresses truths. The investigation of why and with what right we acknowledge a law of logic to be true is, on Frege's view, to be carried out in this language: the reduction of one law of logic to another is a Begriffsschrift proof. Thus for the basic logical laws, logic can give no answer.

On the contemporary view, in contrast, a sentential expression (or formula) of logical notation makes a statement (can be true or false) only relative to an interpretation—another sort of mathematical entity. Consider, for example, the following expression

$$(x)(Fx \vee \sim Fx).$$

On Frege's understanding, even if '*F*' does not actually appear in a quantifier, the above sort of expression is to be understood as expressing a general law—that for every property *F* and object *x*, either *F* holds of *x* or it does not. But on the contemporary view this expression has meaning only on an interpretation that tells us what domain of objects the universal quantifier ranges over and assigns a particular property (or subset of the domain) to '*F*'. Thus the above formula may tell us, on one interpretation, that every natural number is either even or not and, on another interpretation, that every cat is either four-legged or not. The contemporary notion of logical truth (or validity) is understood, also, in terms of interpretation. The above formula is logically valid because it is true under every interpretation. On this conception, logic can always answer the question why and with what right we acknowledge a law of logic to be true. The answer is provided not by a proof from another law of logic, but by a metatheoretic proof: a proof that shows that the expression of the law is true under every interpretation.

One can regard logic, in its contemporary conception, as the basis of all thought (although one need not). Philosophers who hold this view regard the notion of truth under interpretation as a fundamental part of our understanding of the basic rules governing inference and thought. This has significance for our understanding of natural language, as well as logical notation. On this view, a statement of natural language is logically valid just in case it has a regimentation, in logical notation, that is true under every interpretation. Natural language (or at least a cleaned up version of a fragment of natural language) is to be understood as a formal language along with an intended interpretation. Truth, for sentences of natural language, is to be understood as truth under the intended interpretation.

Frege's discussions of language and logic provide important contributions to working this view out. Indeed, much of the contemporary strategy for defining truth under interpretation comes from his writings. One result of his function/argument analysis of statements is that what a complex expression designates is determined by what its constituents designate. Since a sentence is true (or false) just in case it designates the True (or the False), the truth-value of a sentence is determined by what is designated by its parts. Although many believe that it was a mistake on Frege's part to take sentences as names of truth-values, the view that whether a sentence is true is determined by what is designated by its constituents has constituted a starting-point for most contemporary accounts of truth. Frege's distinction between the sense of an expression and what it designates also remains important on this view. An account of truth for natural language requires an account of the intended interpretation. For example, in order to determine the contribution a particular proper name makes to the truth-value of a sentence in which it appears, one needs to know what object it designates on the intended interpretation. One of the important controversies in contemporary philosophical thought about language is whether a proper name hooks onto the object it designates directly or via something like Frege's notion of sense.

Because so many central contributions to this sort of philosophical account of language have their origin in Frege's writings, many philosophers believe that Frege himself took a theory of the workings of language to be a crucial part of the story of why and with what right the laws of logic are true. Yet this view is controversial. One reason is that Frege himself never explicitly says anything of the sort. Another is that, if we take Frege to have held this view we are forced to conclude that his discussion of the basis of arithmetic in *Basic Laws* is either dishonest or hopelessly confused. Frege purports to be providing gapless Begriffsschrift proofs of truths of arithmetic from the primitive truths on which they depend. The premisses of Frege's Begriffsschrift proofs, however, are not truths about language (or Begriffsschrift and its intended interpretation). They are logical laws stated in Begriffsschrift. But if the primitive truths are truths about language then his explicitly stated standards require him to state these truths in Begriffsschrift and prove the logical laws from them. He could have done this, but he did not. It is difficult to imagine that this was a deliberate omission, but it is just as difficult to imagine that he did not notice that his standards required him to offer such proofs. These considerations have led some philosophers to interpret Frege's discussions of language differently: as part of the propaedeutic to, rather than the foundation of, his logic. On this view, these discussions are simply designed to get his readers to understand how his logical system works. The fundamental truths that underlie all thought consist in their entirety in the logical laws expressed in Begriffsschrift.

Although there are puzzles about how to read Frege's work, the controversies are driven, not just by the existence of such puzzles, but by the conviction, on the part of almost everyone involved, that Frege has much to teach us about the issues that concerned him. Frege's work attracted only a small audience in his lifetime. But in the years since, his influence on contemporary philosophy, especially on thought about language and logic, has become ubiquitous.

# *Suggestions for further reading*

## Frege's Writings:

Frege's more influential writings (and most of the writings cited in this book) can be found in:

*The Frege Reader*, ed. M. Beaney (Blackwell Publishers, 1997).

Although large parts of *Basic Laws* have never been translated, English translations of the full texts of his other writings (published and unpublished) as well as translations of parts of *Basic Laws* can be found in the following:

*Basic Laws of Arithmetic: Exposition of the System*, ed. and trans. M. Furth (University of California Press, 1964).

*Collected Papers on Mathematics, Logic, and Philosophy*, ed. B. McGuinness, trans. M. Black *et al.* (Basil Blackwell, 1984).

*Foundations of Arithmetic*, trans. J. L. Austin, 2nd edn. (Basil Blackwell, 1980).

*Philosophical and Mathematical Correspondence*, ed. G. Gottfried, H. Hermes, F. Kambartel, C. Thiel, A Veraart; abridged from German by B. McGuinness, trans. H. Kaal (University of Chicago Press, 1980).

*Posthumous Writings*, ed. H. Hermes, F. Kambartel, F. Kaulbach; trans. P. Long and R. White (University of Chicago Press, 1979).

*Translations from the Philosophical Writings of Gottlob Frege*, ed. P. Geach and M. Black, 3rd edn., (Basil Blackwell, 1980).

There is a great deal of secondary literature on Frege's work and, also, a great deal of controversy about how his work should be understood. Listed below are a few book-length treatments, as well as one collection of shorter articles, that represent a range of different interpretations.

Michael Beaney, *Frege Making Sense* (Duckworth, 1996), an account that focuses on Frege's notion of sense, with some discussion of the historical mathematical and logical background.

Wolfgang Carl, *Frege's Theory of Sense and Reference* (Cambridge University Press, 1994), an account that identifies Frege's notions of sense and *Bedeutung* as playing a primarily epistemological role.

Michael Dummett, *Frege: Philosophy of Language* (Duckworth, 1973), a highly influential account that antedates and, to a large extent, sparked current exegetical controversies. On this account, Frege identifies the theory of meaning as the fundamental part of philosophy which underlies all others.

Michael Dummett, *Frege: Philosophy of Mathematics* (Duckworth, 1991), an account of the role played by *Foundations* and *Basic Laws* in Frege's project of showing that the truths of arithmetic are analytic.

Thomas Ricketts (ed.), *The Cambridge Companion to Frege* (Cambridge, forthcoming), a collection of articles that includes discussions of most of the central issues of concern to contemporary readers of Frege.

Nathan Salmon, *Frege's Puzzle* (MIT, 1986), an example of work in the philosophy of language in which Frege's concerns are taken to be the same as those that move philosophers today.

Hans Sluga, *Gottlob Frege* (Routledge, 1980), an account of the historical origin of Frege's concerns, with emphasis on the philosophical background.

Joan Weiner, *Frege in Perspective* (Cornell, 1990), an exposition and defence of the interpretation that has been presented in this book.

# Index